T0319259

Crack Analysis in Structural Concrete

Crack Analysis in Structural Concrete
Theory and Applications

Zihai Shi

AMSTERDAM • BOSTON • HEIDELBERG • LONDON
NEW YORK • OXFORD • PARIS • SAN DIEGO
SAN FRANCISCO • SINGAPORE • SYDNEY • TOKYO

Butterworth-Heinemann is an imprint of Elsevier

Butterworth-Heinemann is an imprint of Elsevier
30 Corporate Drive, Suite 400
Burlington, MA 01803, USA

Linacre House, Jordan Hill, Oxford, OX2 8DP, UK

Library of Congress Cataloging-in-Publication Data
Application submitted.

British Library Cataloguing-in-Publication Data
A catalogue record for this book is available from the British Library.

ISBN: 978-0-7506-8446-0

For information on all Butterworth–Heinemann publications,
visit our web site at: *www.elsevierdirect.com*

Printed in the United States of America
09 10 11 12 13   10 9 8 7 6 5 4 3 2 1

To my wife and daughter

Contents

Preface

This book is an outgrowth of my research in the broad field of fracture mechanics over a period of twenty years, the past fifteen years of which I have spent focusing on a subbranch of the discipline—that is, fracture mechanics of concrete. My late decision to focus on this field of study was motivated by two factors, namely, a surging demand for crack analysis in structural concrete and a keen personal interest in the subject. Compared with other mature engineering disciplines, fracture mechanics of concrete is still a developing field that is wonderfully rich in scope and diversity and full of challenging issues to be studied.

In recent years a wide range of models and applications have been proposed for crack analysis, and an impressive array of useful information has been accumulated. As a result, the theoretical basis of the discipline has been strengthened; a number of fundamental issues solved; and the range of applications widened. As the subject is approaching its early stage of maturity, it is imperative for students to learn the fundamental theoretical advances that have been made, and engineers need to familiarize themselves with newly developed numerical solution techniques.

I have written this book to summarize the recent theoretical advances in the computational fracture mechanics of concrete, especially regarding the discrete approach to multiple-crack analysis and mixed-mode fracture. The extension of the Fictitious Crack Model (FCM) to address these problems has greatly expanded the range of crack analysis in structural concrete. The book begins with a brief introduction to the fundamental theories of linear elastic fracture mechanics and nonlinear fracture mechanics of concrete. Then, after addressing the issue of stress singularity in numerical modeling and introducing some basic modeling techniques, the Extended Fictitious Crack Model (EFCM) for multiple-crack analysis is explained with numerical application examples. This theoretical model is then used to study two important issues in fracture mechanics: (1) crack interaction and localization and (2) failure modes and maximum loads. The EFCM is subsequently reformulated to include the shear transfer mechanism on crack surfaces and the method is used to study experimental problems. Following these theoretical developments, an application example in tunnel engineering is discussed, which shows how the EFCM can be built into a pseudoshell model for crack analysis of tunnel linings that takes the earth–tunnel interaction into account. Because the book is written both for students and practicing engineers, an effort has been made to present a balanced mixture of theory, experiment, and application.

The companion website for the book contains the source code of two computer programs developed by the author's team, with which the numerical solutions of numerous sample problems discussed can be verified. The purpose of publishing these programs is threefold. First, students can use them to resolve some of the sample problems as exercises to gain a more in-depth understanding of the subject. Second, practicing engineers can use them to solve real engineering problems as this book fully demonstrates. Third, research scientists can use or modify them for specific research purposes. Although great effort has been made to verify the programs, the user must be solely responsible for their performance in practice.

ACKNOWLEDGEMENTS

This book was written with help from a number of people. I am indebted to all the members of my team, present and former, who worked diligently to make valuable contributions. The programs were written by Masaki Suzuki and Masaaki Nakano under my supervision and were tested by Yukari Nakamura for this publication. Numerical computations of sample problems were carried out by Masaki Suzuki, Masaaki Nakano, Kazuhiro Yamakawa, and Sadanori Matsuda. Figures were drawn by Hiroko Okabe, whose superb work exceeded my expectations. Masaaki Nakano greatly assisted me in furnishing annotations for the programs and in preparing the materials for Chapters 9 and 10. As the author of this book, however, I must take sole responsibility for any negligence or errors that may remain in it or the programs.

I am also indebted to many people and organizations for their direct or indirect assistance. A special expression of gratitude is extended to the following:

- Dr. Tamotsu Yoshida and Hiroshi Tanaka, the former and present directors of the Research and Development Center of Nippon Koei Co. Ltd., and Dr. Shu Takahashi, the center's general manager, for their firm support for this project that made my work of writing much easier
- Professor Masayasu Ohtsu of Kumamoto University for helpful discussions about a number of issues over the years and for his stimulating influence
- Noriaki Yoshida for his kind assistance in obtaining permission from clients for using some parts of previous project reports, which were the work of my team
- Tokyo Electric Power Company for granting me permission to use some portion of the project reports on the maintenance and management system of its aging waterway tunnels as application examples.

I wish to express my appreciation to the professional societies, book publishers, and technical journals that have given me permission to use copyrighted material in the preparation of this book. I am especially grateful to Professor D. A. Hordijk for his generous consent for me to reuse some important experimental results of his PhD thesis.

I also wish to thank the editorial staff of Elsevier, in particular David Sleeman, Alex Hollingsworth, Lanh Te, Kenneth P. McCombs, Marilyn E. Rash, and Maria Alonso for their diligent efforts in overseeing the publication of the book. The idea of writing it was stimulated by an e-mail from David Sleeman in the summer of 2005; that cordial invitation to write a book on material science made me realize the need for new textbooks to reflect recent advances in computational fracture mechanics of concrete (until then I had never thought of authoring a book by myself). Without his enthusiastic encouragement to put pen to paper (or fingers to keyboard), this book probably would not be here.

Finally, I am thankful to my wife, Xiaoping, and my daughter, Xijia, for being a constant source of happiness, inspiration, and encouragement.

Introduction

1.1 AIMS OF THE BOOK

As research scientists and practicing engineers working in the field of structural engineering and whose responsibilities are almost exclusively confined to analyzing cracked concrete structures for the safety evaluation and renovation design of today's aging infrastructure, we have been studying cracks in various concrete structures through crack analysis for more than a decade. The structures that we have frequently encountered for crack diagnosis include tunnels, dams, bridges, sewage pipes, and so forth, and we need to clarify the various mechanisms by which cracks occur and evaluate the damaging effects that these cracks inflict on these structures. Crack analysis is indispensable in answering these questions, and it is a fascinating field of study in structural concrete. The peculiar nonlinearity that occurs in the fracture process zone (FPZ) ahead of an open crack makes the theory distinctly different from the whole range of problems in classical continuum mechanics. In addition, the frequent encounter of multiple cracks and mixed-mode fracture in real situation problems that require creative approaches highlights the challenging feature of our work, which compels one to advance fracture mechanics of concrete further and to expand its applications wider.

Let us be specific. In Photo 1.1 a large longitudinal crack is shown in the right portion of the arch of an aging highway tunnel. Field surveys showed that the crack-mouth-opening displacement (CMOD) had reached a maximum width of 7 mm, and circumstantial evidence also suggested the existence of another crack in the left portion of the arch from the outer surface of the tunnel lining. Field investigations found that a narrow fault (50 cm in thickness) composed of class D rock mass ran through the tunnel cross section from upper right to lower left direction. A loosening zone with a depth of more than 3 m composed mainly of class CL rock mass was found along the fault. A schematic illustration of the situation is shown in Figure 1.1. Our task may be simple to express, but it is difficult to solve—that is, to determine, through crack analysis, the pressure loads exerted on the tunnel lining by the loosening zone. This information could provide the much needed design load for remedial works to be carried out on the tunnel linings to stabilize the crack and ensure the safety of the tunnel.

Now we have a crack analysis problem with two discrete cracks, stemming from an actual engineering situation. Let us specify our solution strategy. First, crack analysis will be carried out using a pseudoshell model to determine the cross-sectional deformation of the tunnel lining

PHOTO 1.1 A highway tunnel with a large longitudinal crack in the right arch area.

as the CMOD of the inner lining crack reaches 7 mm under the earth pressures. Next, assuming a simple elastic loosening zone model and that the ground deformation of the loosening zone is equivalent to the cross-sectional deformation of the tunnel lining at the crack, then the depth of the loosening zone can be determined through iterative computations. Once the size of the loosening zone is known, the pressure loads acting on the tunnel lining can be calculated from its gravity loads. The flow chart for this numerical computation is shown in Figure 1.2.

Before starting this numerical analysis, we must choose a numerical method for modeling cracks. In fracture mechanics of concrete, there are two generally accepted computational theories used in the finite element method (FEM) analysis to represent cracking in structural concrete: the smeared crack approach (Rashid, 1968) and the discrete crack approach (Ngo and Scordelis, 1967; Nilson, 1967; Nilson, 1968; Hillerborg et al., 1976). The smeared crack approach treats the cracked solid as a continuum and represents cracks by changing the constitutive relations of the finite elements. The wisdom of this approach lies in its full exploitation of the fundamental concepts of the FEM, which can be summarized as (1) subdivide a problem into elements, (2) analyze this problem through these elements, and (3) reassemble these elements into the whole to obtain the solution to the original problem.

In some sense, the smeared crack approach may be considered as a straightforward way to express cracks with the FEM, because the stress-strain relations of the elements can be easily altered to reflect the effects of cracking, which is a very convenient feature not only for static crack analysis but also for dynamic crack analysis, such as simulations of dynamic crack propagation during earthquakes. A different modeling concept is adopted in the discrete crack approach, in which a crack is treated as a geometric entity and the FPZ forms a part of the boundary condition, including the geometric shapes and cohesive forces acting on the crack surface.

From a macroscopic-material point of view, this is the most accurate physical model to study cracks because the approach reflects most closely the physical reality of the cracked concrete.

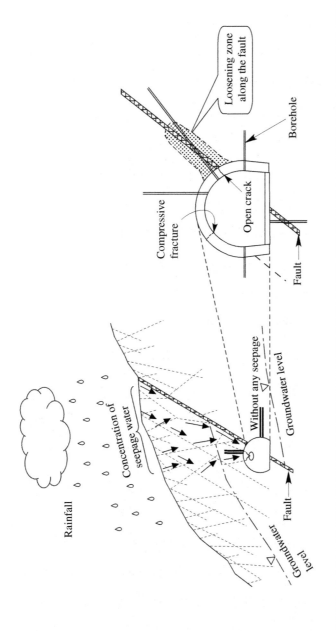

FIGURE 1.1 Schematic cross section of a highway tunnel with large, open cracks based on the results of detailed field investigations.

3

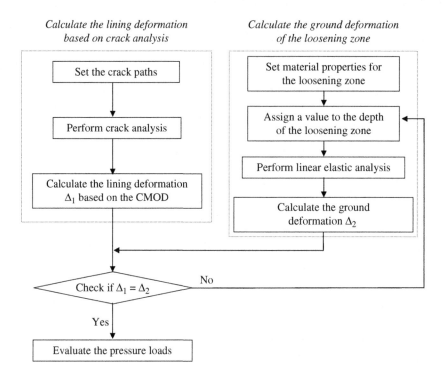

Calculate the lining deformation based on crack analysis   *Calculate the ground deformation of the loosening zone*

FIGURE 1.2 Evaluation of the pressure loads based on a crack analysis of the tunnel lining.

It seems that these two modeling concepts emerged naturally in the 1960s through the great efforts of engineers attempting to analyze concrete structures using the FEM, following the advent of computers in the 1940s and the subsequent rapid development of the FEM in the 1950s. Due to its continuum assumption for cracked concrete materials, the smeared crack approach is computationally much more convenient than the discrete crack approach.

The method has been widely accepted in practice as one of the most effective means for crack analysis in concrete structures to predict general structural behaviors. However, the limitations of this approach are just as obvious as its merits. In principle, the approach can only predict cracking behavior approximately, and the continuum assumption makes it impossible to obtain any specific information related to the crack-opening displacement (COD) of any individual crack. Therefore, the smeared crack approach is unfit for our present task. On the other hand, the modeling concept of the discrete crack approach makes it possible to analyze cracking behavior in a concrete structure as physically accurately as possible, including the crack path and the CMOD. For this reason, the discrete crack approach, which is the focus of this book, will be chosen to solve our present problem.

Over the years, various other analytical concepts and modeling techniques have also been proposed for specific research purposes, such as the microplane theory by Bazant and Ozbolt (1990), the particle model by Bazant et al. (1990), and the lattice model by van Mier (1997), to name but a few, reflecting the complex and diverse mechanisms of various fracture phenomena in concrete structures. In practice, applications of these models are limited due to the various unique material assumptions they adopt. It is unrealistic to cover such diverse topics for crack analysis in this book, and references for some of these works can be found in the ACI 446 report (1997).

Next, the specific features of a numerical model based on the discrete crack approach should be discussed. Earlier attempts in simulating crack propagation in concrete did not include the tension-softening phenomenon in the FPZ, and both of those terms (i.e., the tension-softening and the FPZ) were unknown concepts then. These two ideas were first proposed by Hillerborg, Modeer and Petersson in 1976 as both the physical realities in cracked concrete and important constitutive details of a numerical model for studying crack formation and crack growth in concrete, based on the experimental evidence of tension tests. As is now known, the FPZ and the tension-softening relation have become the fundamental concepts of fracture mechanics of concrete. In some respects, the fictitious crack model (FCM) proposed in their pioneering work may be considered as an application of two important mathematical models for crack tip modeling, proposed by Barenblatt (1959, 1962) for brittle fracture and by Dugdale (1960) for plastic yielding in steel, respectively. Figure 1.3 illustrates the terminology and concepts associated with the FCM.

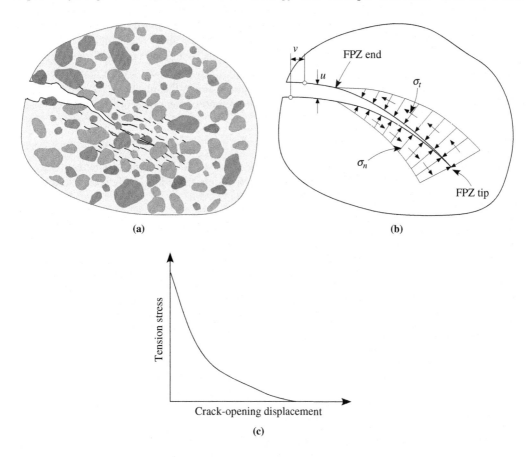

FIGURE 1.3 The fictitious crack model: (a) a crack in concrete, (b) a fictious crack, and (c) tension-softening relation (Hillerborg et al., 1976).

This model assumes that the FPZ is long and infinitesimally narrow, which "in reality corresponds to a microcracked zone with some remaining ligaments for stress transfer" (Hillerborg et al., 1976). The FCM is now widely accepted as one of the simplest and most accurate nonlinear discrete fracture mechanics models applicable to fracture of concrete structures.

Carrying out a crack analysis using the FCM is not a simple task. At any given load level, a solution is sought not only for the stress and strain fields but also for the geometric shapes and cohesive forces of the fictitious crack, which itself forms a part of the boundary condition. In general, this constitutes a nonlinear problem with unknown boundaries. Quoting from the important work by Barenblatt (1962):

> In the theory of cracks one must determine from the condition of equilibrium not only the distribution of stresses and strains but also the boundary of the region, in which the solution of the equilibrium equations is constructed.... Likewise, the basic problem in the theory of equilibrium cracks is the determination of the surfaces of cracks when a given load is applied.

For certain types of problems with unknown boundaries, closed-form solutions may exist. The Dugdale model presents a closed-form solution for the extent of plastic yielding under external load. The original problem of yielding at the end of a slit in a sheet was simplified as an infinite plate under uniform tension in the direction perpendicular to an internal cut of total length $2a$. Yielding occurs over the unknown length ρ measured from the end of the cut, as shown in Figure 1.4. With the known constant stress distribution along the plastic zone, a solution is obtained by imposing the finite stress requirement at the end of the crack ($x = a + \rho$):

$$\rho = a\left[\sec\left(\frac{\pi\sigma}{2\sigma_{ys}}\right) - 1\right] \tag{1.1}$$

where σ is the tensile stress at infinity, and σ_{ys} is the yield stress. For crack problems defined by the FCM with a nonlinear tension-softening relation along the FPZ, analytical solutions are unavailable. Such problems must be solved numerically.

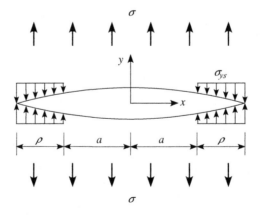

FIGURE 1.4 The Dugdale model.

In their pioneering work, Hillerborg and his coworkers not only established the basic concepts of the FPZ and the tension-softening relation but also specified a numerical analysis procedure for predicting crack propagation in a plain concrete beam with a single crack. To explain their solution strategy, the original problem is shown in Figure 1.5, which was a simple beam subjected to pure bending moment M. In their study, only one crack was allowed to propagate from the midspan, so only one-half of the beam needs to be analyzed due to symmetry. Let us assume that the moment $M = M_0$ produces $\sigma_{31} = f_t$, where σ_{31} is the normal tensile stress at node 31, and f_t is the tensile strength of concrete. This indicates that M_0 is the crack initiation moment. As a result, the

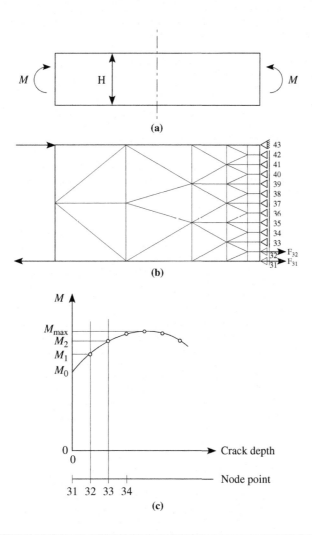

FIGURE 1.5 Hillerborg's original problem with FCM: (a) beam in pure bending, (b) finite element mesh of the beam, and (c) moment-crack depth curve (after Hillerborg et al., 1976).

hinge support at node 31 is removed, and a cohesive force F_{31}, determined from the normal stress versus COD curve—that is, the tension-softening relation—is applied against the crack opening at the same node, as shown in Figure 1.5b. At this point, the previous boundary condition has been changed partially to accommodate the crack propagation. Then, the next step of the analysis is carried out with the crack tip now at node 32, and the beam is subjected to a new load region of the moment force M and the cohesive force F_{31}. After the moment M_1, which leads to $\sigma_{32} = f_t$ at node 32, is determined, the hinge support at node 32 is removed and another cohesive force, F_{32}, is introduced. With increasing M, the values of F_{31} and F_{32} may change due to the increase of the COD at each node. Proceeding with similar calculations by changing the moment and the cohesive forces, the crack propagation in the beam can be predicted. The relation between the applied moment and the crack length is shown in Figure 1.5c.

Based on the preceding discussion, the computational procedure can be summarized as follows:

1. Prescribe a new crack path, which forms a part of the new geometric boundary condition.
2. Apply the cohesive forces on the crack surface, which are determined based on the tension-softening relation and become a part of the load region.
3. Carry out the stress analysis with the new boundary condition as defined in steps 1 and 2.
4. Repeat the first three steps until structural failure.

In the step-by-step numerical analysis, the direction of the next-step crack propagation is determined based on the results of stress analysis at the present load step. The crack propagates in the direction normal to the maximum principal tensile stress at the tip of the crack.

For a prescribed incremental crack growth, the distribution of the cohesive stresses at the fictitious crack can be determined based on a method that is now known as the influence function approach. Influence coefficients are used in the superposition of FEM solutions to establish the governing equations for crack propagation. Details of the influence function method will be discussed later in the book. Once all of the unknown boundary conditions are clarified, the original problem is reduced to an ordinary boundary value problem, and its stress solution can be obtained easily. Based on the newly obtained stresses, the next incremental extension of the crack is prescribed, and a new round of computation proceeds.

As an illustrative example, the computational procedure of the FCM is applied to numerically obtain the relations between the extent of plastic yielding and the applied load in Eq. (1.1). Due to symmetry, only one-quarter of the problem in Figure 1.4 is analyzed. As shown in Figure 1.6a, the plastic zone (crack path) is set along the cut, and yielding (crack propagation) takes place when the tensile stress at the tip of the cut reaches the yield stress σ_{ys}. One may assume $\sigma_{(0)}$ as the yield initiation load leading to $\sigma_y^{(1)} = \sigma_{ys}$, where $\sigma_y^{(1)}$ is the normal tensile stress at node 1. Consequently, a yield force F_1 is applied at node 1, which tends to close the crack, as shown in Figure 1.6. The yield force F_1 is determined based on σ_{ys} and the area apportioned to node 1. With the tip of the plastic zone moving to node 2, the next step of the analysis is carried out under a new load region of σ and F_1. After the determination of $\sigma_{(1)}$, which produces $\sigma_y^{(2)} = \sigma_{ys}$ at node 2, another yield force, F_2, is introduced at node 2. It is noted that the values of F_1 and F_2 remain constant as the COD at each node increases with increasing load. Repeating similar calculations by changing the applied load and adding a yield force at the tip of the enlarging plastic zone, the relation between the applied tensile stress and the size of the plastic zone can be obtained, as shown in Figure 1.6b.

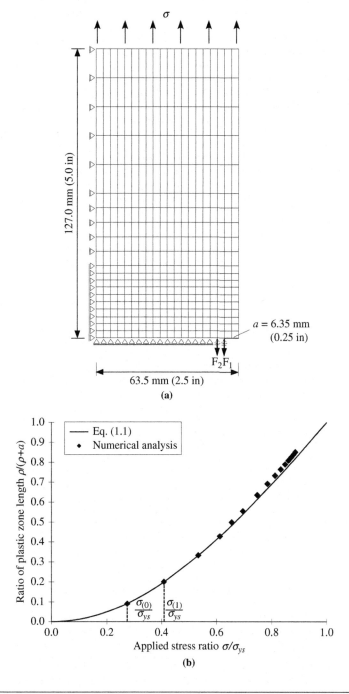

FIGURE 1.6 Numerical analysis of Dugdale's tensile tests on steel panels: (a) FE model for Dugdale's tensile tests of steel panels and (b) relation between load and plastic zone length.

In establishing the FCM, Hillerborg et al. made these two important assertions:

1. The FCM could be applied to multiple-crack problems.
2. The FCM could be applied to mode-II and mode-III shear-type fracture.

It became clear later that neither of these assertions could be achieved by a straightforward extension of the FCM. First, a multiple-crack problem is very different from a single-crack problem in several respects. In a multiple-crack situation, every crack has three possibilities of reaction when subjected to external loads, including crack propagation, crack arrest, and crack closure. This leads to multiple modes of crack propagation that have to be considered in a numerical analysis.

On the other hand, when only a single crack is involved, such as in Hillerborg's beam problem or in Dugdale's plate problem, the mode of reaction for the crack is clear according to the Griffith energy theory (1921, 1924), which states that a crack will propagate if the energy released upon its growth is sufficient to provide all the energy that is required for its propagation. While this single mode of crack propagation greatly simplifies the software logistics for solving a single-crack problem, new solution strategies have to be found for selecting the true cracking behavior from among the many possibilities when multiple cracks are encountered. Second, to extend the FCM to pure mode-II or mode-III fracture seems unrealistic, since the existence of shear-type fracture in reality is questionable (Bazant and Gambarova, 1980).

It is widely known that most of the practical fracture problems are of a mixed-mode nature, including both the opening mode and the shear modes. To combine mode II or mode III with mode I to form a mixed-mode crack based on the FCM, a shear-lag phenomenon in which shear transfer is delayed in the initial stage of crack propagation must be taken into account because the slip (causing frictional forces as shear) on the crack surface can occur only after some finite opening has already been achieved (Bazant and Gambarova, 1980). Facing these challenges, the aims of this book are threefold: to extend the FCM to multiple-crack problems, to extend the FCM to mixed-mode fracture, and to apply these newly developed computational theories to solve several important theoretical and engineering problems.

Finally, let us proceed with our analysis of the cracked tunnel-lining problem. From a structural point of view, a tunnel lining containing through-thickness cracks, as shown in Photo 1.1 (cracks of this scale in a tunnel lining usually penetrate the whole depth of the wall), can hardly be considered as structurally stable without the interactive support by the surrounding rock mass. In the present study, this interaction is simply replaced by a pseudoshell, which supports the tunnel lining by rigidly connecting to it. (Details of the pseudoshell model will be discussed in Chapter 8.)

Figure 1.7 presents the FE mesh of the tunnel lining and a pseudoshell with an elastic modulus equivalent to that of steel. As seen, two small notches were introduced into the tunnel lining at the locations of the cracks: one from the inside of the lining in the right arch and one from the outside of the lining in the left arch. In the vicinity of the outer notch, the connections of the tunnel lining and the pseudoshell were removed to allow a crack to extend from there. To simulate the pressure loads due to the loosening zone, a dummy load was applied to the pseudoshell, as shown in Figure 1.7. A numerical analysis was carried out by gradually increasing the dummy load to propagate the cracks until the CMOD of the inner crack reached 7 mm. The results of the crack analysis are shown in Figure 1.8. As seen, the two cracks were indeed through-thickness cracks. The cross-sectional deformation of the tunnel lining at the inner crack was obtained as 28.3 mm when the CMOD reached 7 mm.

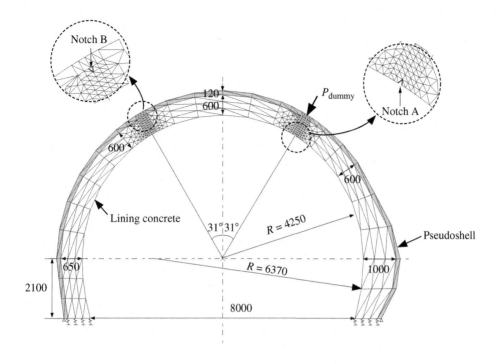

FIGURE 1.7 FE pseudoshell model (dimensions in mm) for the cracked tunnel lining.

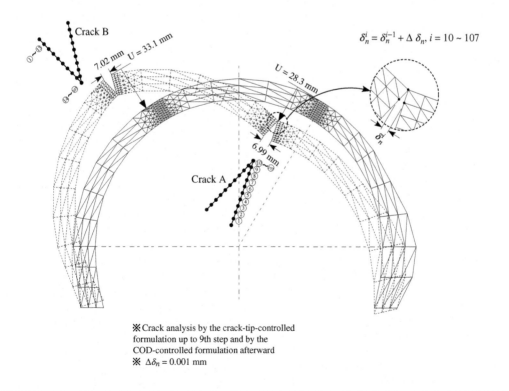

FIGURE 1.8 Numerical results of crack analysis on the tunnel lining.

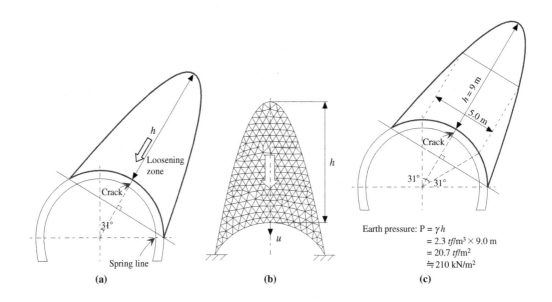

FIGURE 1.9 Evaluation of the earth pressure: (a) loosening zone model, (b) FE model, and (c) numerical results on the depth of loosening zone and the earth pressure.

Next, to estimate the earth pressures, a loosening zone model was assumed with an unknown depth h, as shown in Figure 1.9a. Under the gravity load a loosening zone deforms. Figure 1.9b presents the FE mesh of the assumed loosening zone. Based on linear elastic analysis, it was found that a loosening zone with a depth of 9 m—nearly equivalent to the tunnel diameter—was required to cause the ground deformation equivalent to the cross-sectional deformation obtained by the crack analysis. The obtained size of the loosening zone was in good agreement with the results of detailed field investigations. A loosening zone with a depth of 9 m exerted a pressure load of 210 kN/m² on the tunnel, as shown in Figure 1.9c. Assuming a safety factor of 2, a design load of 450 kN/m² was recommended for remedial work on this tunnel.

1.2 MULTIPLE-CRACK PROBLEMS

In practice, it is rare, if ever, to encounter a crack problem in which only a single crack is involved. Many engineering problems with multiple cracks are simplified as single-crack problems mainly because of the lack of effective means to solve the original problem and sometimes due to the desire for a quick and simple solution based on the safe design consideration. In general, single-crack problems are confined to laboratory test conditions, and these tests are often conducted on plain concrete beams to obtain fracture parameters such as the mode-I fracture energy G_F, to verify new computational theories or to study certain fracture behavior under specific load conditions, and so on.

The Griffith energy theory (1921, 1924) is the basis for understanding fracture phenomena, as it effectually interprets crack propagation as an inevitable process of energy transfer between the strain energy of an elastic body in the equilibrium states and the fracture energy required for creating a new crack surface so as to achieve a state of minimum potential energy for the body at a given load level. To propagate a crack, the load level has to be raised to deform the body until the strain energy stored in the body is sufficient to supply the required fracture energy as the tip stress of the crack reaches a critical stress criterion (such as the maximum principal stress criterion).

Griffith established his energy theory while studying glass, and he therefore assumed that for brittle materials, the fracture energy was composed of the surface energy that was required for creating new crack surfaces. It is now known that for concrete, an FPZ exists at the tip of an open crack, which is a transient zone between the open crack and the undamaged continuous material outside the FPZ (see Figure 1.3a and b). In the FCM the inelastic material behavior of the FPZ is simplified as and represented by a normal stress versus COD curve—that is, the tension-softening relation (see Figure 1.3c). During crack propagation, the resistance of the cohesive stresses of the fictitious crack against crack opening has to be overcome, and the energy consumed in this process is the fracture energy required for creating a new crack surface.

The Griffith theory explains why crack propagation is inevitable when an elastic body is subjected to an increasing tension load. For a single-crack problem this means that the crack must be an active crack. Hence, in a numerical analysis using the FCM, the future crack path that will form a part of the new boundary condition for the next-step crack analysis can be determined without any ambiguity. In a multiple-crack situation, every crack possesses three possibilities of motion—including crack propagation, crack arrest, and crack closure—and combinations of these modes among all the cracks involved will lead to a great number of possibilities for the next-step crack path prediction.

Consider a two-crack problem as an example. Figure 1.10 shows an infinite plate of unit thickness with two arbitrarily positioned interacting cracks, with the lengths of the cracks A1 and A2, respectively. The plate is stressed to allow crack propagation. Figure 1.10a illustrates all of the five possible crack patterns or cracking modes, such as the single-crack propagation modes in Cases 1 and 2, the simultaneous-crack propagation pattern in Case 3, and the crack propagation accompanied by crack closure scenarios in Cases 4 and 5. Since material damage due to fracture is irreversible, the closure of a crack here simply means the closing of the crack surfaces under local compressive stresses, without reversing the damaged material properties. The load-displacement relations are given in Figure 1.10b for the fixed-end condition. Before the crack growth, the elastic energy contained in the plate is represented by the area OAB. After the crack extension, the load will drop to C under the fixed-end condition, and the stiffness of the plate will decrease to the line OC. The new elastic energy content is given by the area OCB.

According to the Griffith energy theory, the elastic energy release represented by the area OAC is consumed as the fracture energy to create a new crack surface ΔA_1 or ΔA_2, or both. Although the Griffith theory ensures that crack propagation will take place in one of the five possible cracking modes shown in Figure 1.10a, it cannot specify which one. As a result, the softened stiffness represented by line OC in Figure 1.10b can be caused by any one of them, whereas, according to Barenblatt (1962), "cracks, whose surface also constitutes a part of the body boundary, can expand a good deal even with a small increase of the load to which the body is subjected."

Since with each cracking mode a new boundary value problem forms, it is obvious that the crack analysis of a multiple-crack problem must involve multiple solutions. As will be explained in Chapter 4, a final solution can be obtained by eliminating unrealistic cracking behaviors that would lead to only invalid solutions and by identifying the true cracking mode from the other feasible cracking modes based on a minimum load criterion (Shi et al., 2001). These solution concepts are the basis of the extended fictitious crack model (EFCM) that will be discussed in Chapter 4.

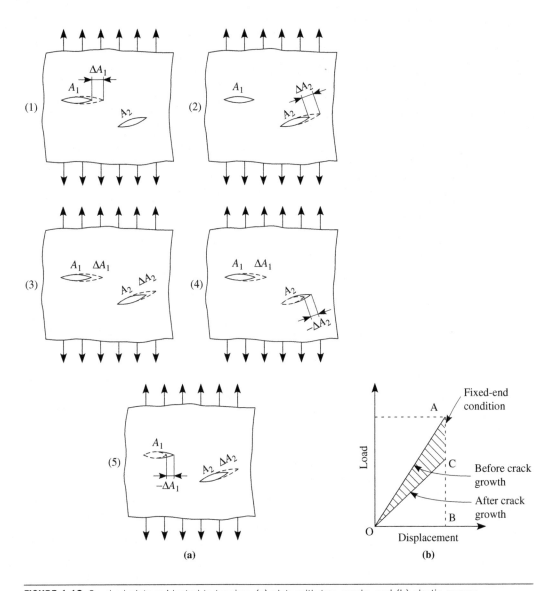

FIGURE 1.10 Cracked plate subjected to tension: (a) plate with two cracks and (b) elastic energy.

One might think the solution procedures just described are unnecessary, since all the information on crack propagation seems to be contained in the strain and stress fields or, more precisely, in the tip stresses of the existing cracks or notches obtained by linear elastic analysis. Based on this conjecture, the tip stress at each existing crack or notch is checked, and if the tensile strength of concrete is reached at a certain point, a crack is deemed to propagate from there. This solution concept is questionable for solving multiple-crack problems.

To begin with, the whole theory of linear elasticity is built on the small deformation assumption, which permits the boundary conditions at the surface of the unstrained body to be satisfied because any deformation occurring to the original geometric shapes is stipulated as small and negligible. However, crack extension will cause distinct changes to the present boundary condition (not only the change of the geometric shapes of the body due to the newly created crack surfaces but also the change of the force boundary condition because cohesive forces appear on these newly created crack surfaces), and these changes will have to be taken into account to form new boundary conditions under which a new stress state is to be calculated.

Obviously, the true solution to the problem is among the solutions obtained with the boundary conditions derived from relevant cracking modes. Due to the small deformation assumption in the linear elastic theory, the present strain-and-stress fields do not contain sufficient information on the next-step cracking behaviors for multiple cracks, which must be supplemented from the cracking mode analysis as discussed in the two-crack problem of Figure 1.10. It should be noted that in multiple-crack problems that involve little crack interaction, it is indeed possible to identify an active crack by just checking the tip stresses at the existing cracks or notches. This is almost self-evident, however, because in these situations there exists effectively only one stress concentration point from which a crack will extend when sufficient load is applied.

To illustrate this point, Figure 1.11 schematically shows three situations corresponding to a single-crack case and two multiple-crack cases—one with little crack interaction and one with

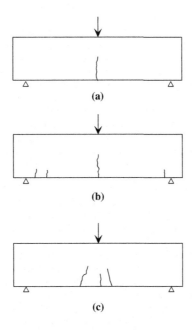

FIGURE 1.11 Propagating cracks and nonpropagating cracks: (a) single-crack problem, (b) multiple-crack problem with little crack interaction, and (c) multiple-crack problem with lots of crack interaction.

strong crack interaction. As with most multiple-crack problems, crack interactions are the source of some crucial fracture mechanisms for causing a variety of cracking behaviors, which can only be predicted accurately in a numerical analysis when adequate cracking modes are taken into consideration, as just discussed.

Among the many approximate methods that have been proposed for analyzing multiple-crack problems in concrete structures, discrete modeling methods using interface elements have been actively explored by the research community (Ingraffea et al., 1984; Ingraffea and Saouma, 1984; Ingraffea and Gerstle, 1985; Rots and Schellekens, 1990; Gerstle and Xie, 1992; Reich et al., 1993; Cervenka, 1994; Shah et al., 1995; Xie and Gerstle, 1995; ACI 446 report, 1997). This approach may resemble the FCM in some respects, but there are fundamental theoretical differences between the two that are sometimes overlooked.

In this approach, the FPZ is often modeled by using zero-thickness interface elements that are formulated with the normal and shear stresses and relative displacements across the interfaces as constitutive variables. Separation of the interfaces is stipulated by imposing a constitutive relation between the cohesive tractions connecting the interfaces and the displacement discontinuities across the interfaces. The obtained numerical results show discrete crack configurations similar to those obtained by the FCM. As with the smeared crack approach, however, the judgment for crack propagation is based on the current stress state. The applications of the interface elements are wide, ranging from single-crack problems to multiple-crack problems, from static crack propagation to dynamic crack propagation. Nevertheless, one must keep in mind the limitations of this approach in solving multiple-crack problems. Due to its incomplete judgment on crack propagation, which may well exclude some valid cracking modes from the solution, accurate predictions for complicated cracking behaviors may sometimes be unachievable.

1.3 MIXED-MODE CRACK PROBLEMS

In making their assertions that the FCM may also be applied to mode-II and model-III shear-type fracture, Hillerborg and his coworkers (1976) apparently envisioned certain types of shear transfer rules similar to the tensile stress versus COD relations that they had proposed for the mode-I crack, as shown in Figure 1.12. These relations, leaving their specific details aside, present well-accepted concepts on various material damage phenomena such as yielding for metals in Figure 1.12a, and decrease of the tensile strength of concrete due to cracking in Figure 1.12c.

This reflects the fact that a mode-I crack is a relatively clear type of fracture, probably because its macroscopic damaging effect on the strength of the material is consistent with our atomic view of fracture regarding the forced separation of two atoms that are bounded together by atomic attractions. On the contrary, the shear-type fracture of both mode II and mode III is complex, and the independent existence of these modes to mode I in structural concrete is even questionable. Since most of the practical fracture problems are of a mixed-mode nature, involving mode I and mode II, the research community has long focused on mixed-mode fracture.

To clarify the shear transfer mechanism due to aggregate interlocking at crack surfaces in structural concrete, pioneering experimental investigations were carried out by a great number of researchers in the 1960s and the 1970s. Based on the experimental results obtained by Paulay

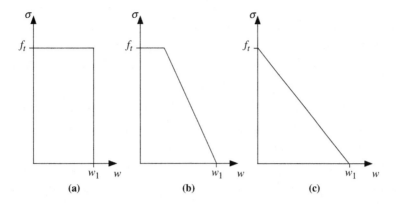

FIGURE 1.12 Examples of the tension-softening relation proposed by Hillerborg et al. (after Hillerborg et al., 1976).

and Loeber (1974), Bazant and Gambarova (1980) conducted a thorough theoretical analysis on the stress and displacement discontinuity relations on rough cracks and published a landmark paper in 1980 that presented insights and guidelines for numerical modeling of mixed-mode fracture. In that paper they said the following:

> *Thus, the first displacement on the rough crack must be normal, and the slip can occur only after some finite opening has already been achieved. This condition must be carefully followed in numerical calculations. . . . Continuous cracks in concrete must propagate in such a direction that the displacement field near the crack tip be purely of Mode I (opening) type (i.e., the Mode II field cannot exist).*

These principles have proven fundamental for understanding the nature of the phenomenon as well as for numerical modeling of the shear transfer mechanism.

Given specific tension-softening and shear transfer laws, a numerical formulation of a mixed-mode fracture in the FCM poses no particular conceptual difficulties as compared with the mode-I formulation (Shi, 2004). For crack propagation, the maximum principal stress criterion is used because the mode-I condition is dominant at the tip of a mixed-mode crack, based on the preceding analysis by Bazant and Gambarova. As the tip stress reaches the tensile strength of concrete, a mixed-mode crack propagates. It should be noted that the inclusion of shear transfer in a numerical formulation might significantly affect cracking behavior. Within the limited studies conducted by the author's team, it seems that shear transfer on crack surfaces has its greatest influence on the postpeak structural response and the crack path but not very much on the peak load. In general, the postpeak load-deformation curve becomes less brittle as compared with the response curve under the mode-I condition, reflecting the increased energy consumption in mixed-mode fracture.

Similarly, due to the existence of frictional forces on the crack surfaces, the curvature in a mixed-mode crack path is generally more restrained or gentler than the corresponding mode-I crack path when the shear transfer mechanism is eliminated from the crack surface. These distinctive features associated with a mixed-mode crack highlight the significance of the inclusion of shear transfer in crack analysis in order to accurately predict the cracking behaviors when a mixed-mode fracture problem is encountered. Furthermore, a mixed-mode formulation reveals new potentialities for advanced material characterization. Quantities that have been difficult to obtain or even unmeasurable in the past, such as the shear strength of concrete and the mode-II fracture energy, may now be quantified within the framework of a mixed-mode fracture model by combined testing and numerical simulation. These topics will be elaborated on in Chapter 7.

1.4 CRACK INTERACTION AND LOCALIZATION

Crack interactions among multiple cracks can have a strong influence on the local stress field and crack driving force for a given crack, eventually leading to damage localization by the forced closure of some cracks or the coalescence of others. Many studies have examined the interaction effects of multiple cracks in the fracturing process of concrete (Bazant and Wahab, 1980; Ingraffea et al., 1984; Barpi and Valente, 1998; Bazant and Planas, 1998; Carpinteri and Monetto, 1999). In the continuum damage models, these effects are reflected through continuum smearing of microcracks in the stress-and-strain fields, which leads to a nonlocal continuum in which the stress at a point depends also on the strains in the neighborhood of that point. This approach provides an effective means for investigating the gross effect of crack interaction on the general cracking behavior.

On the other hand, individual cracking behaviors have to be captured by using discrete modeling methods that allow the interaction of multiple cracks to be studied most straightforwardly. Using the EFCM, an explicit mathematical expression for crack interaction can be deduced based on the classification of the stress components of the crack driving force at each crack (Shi et al., 2004). Known as the *coefficient* of crack interaction, this physical term enables the interaction effect to be quantified for each individual crack and various cracking behaviors and fracture mechanisms to be clarified based on this quantified information.

General questions related to crack interaction may be summarized as follows:

1. In the process of crack propagation, why are some cracks active, while others are not?
2. Why is crack/damage localization inevitable?
3. Why does crack localization begin early in some cases and late in others?
4. Under what circumstances do cracks coalesce?
5. Is there any effective means for identifying the potential critical cracks from a group of cracks in concrete structures?

Many other questions can be added to this list. Topics related to crack interaction and localization will be discussed in Chapter 5.

1.5 FAILURE MODE AND THE MAXIMUM LOAD

In a numerical study of the load-carrying capacity of notched concrete beams, a strong dependence of the maximum load on the failure mode was reported (Shi and Suzuki, 2004). The loading conditions and notch arrangements of the simple beams in the original study are illustrated in Figure 1.13. As shown, among the three notches introduced into the beam notches, A and B were kept at a constant size of 10 mm, while notch C was assigned various sizes to study the relations

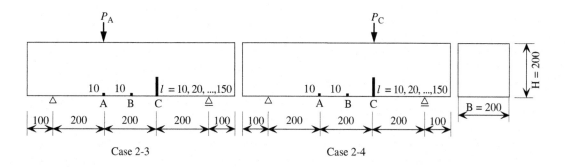

Case 2-3

Case 2-4

FIGURE 1.13 Numerical models (dimensions in mm) of eccentric loading tests (after Shi and Suzuki, 2004).

between the maximum loads and failure modes under eccentric loading. The obtained relations are shown in Figure 1.14, which contains two curves. When the eccentric load was applied at notch C, a monotonically decreasing relation between the peak load and the size of notch C was obtained, and the dominating crack for beam failure was shown to invariably develop from notch C.

On the other hand, when the eccentric load was applied at notch A, the obtained maximum load seemed to be unaffected by the enlargement of notch C until it reached a critical value, beyond which the peak load decreased quickly as the size of notch C increased. It was shown that two failure modes were involved in the latter case. Before reaching the threshold value of notch C, the dominating crack for the beam failure originated from notch A; beyond that point it developed from notch C. As clearly revealed by this study, the change of cracking behavior and failure mode may result in a significant reduction of the load-carrying capacity of the simple beam. This fact may have significant implications in clarifying the fatigue mechanisms of certain engineering materials.

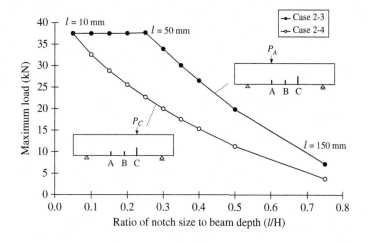

FIGURE 1.14 Relationship between maximum load and the ratio of notch size to beam depth (after Shi and Suzuki, 2004).

In studies of metal fatigue, it has long been known that the fatigue strength of a test specimen is not affected by introducing artificial micro holes into the specimen unless the size of these artificial defects exceeds a critical value, beyond which fatigue strength decreases significantly (Murakami, 1993). In general, initial defects exist in all structural members; only the degree of imperfection varies. Under cyclic loading, the material weakening process of a structural member inevitably involves multiple cracking originated from some of these spatially distributed initial imperfections. As a threshold value in terms of, for example, a critical crack length or a critical crack density (such as the maximum number of cracks in a certain location) is approached, unexpected cracking behaviors may abruptly emerge and replace the previous ones, causing a sudden degradation in material strength. Obviously, this process can repeat itself until the remaining material strength can no longer sustain the level of stress produced by the design load, leading to abrupt structural failure. This important topic will be explored in Chapter 6.

1.6 OUTLINE OF THIS BOOK

This book focuses on the latest developments in computational theories on multiple-crack analysis and mixed-mode fracture in structural concrete and the application of these theories to solve important engineering problems. Chapters 1 to 3 introduce the fundamental theories of fracture mechanics. Chapter 1 is a general introduction, Chapter 2 covers the fundamentals of linear elastic fracture mechanics and nonlinear fracture mechanics of concrete, and Chapter 3 examines the fictitious crack model and related issues in its numerical implementation.

Chapters 4 to 7 present computational theories on multiple cracks and mixed-mode fracture. Chapter 4 introduces the computational theory for analyzing multiple discrete cracks of the mode-I type (EFCM), which is then applied to study two important issues in fracture mechanics: crack interaction and localization (Chapter 5) and failure modes and maximum loads (Chapter 6). Chapter 7 discusses the issue of mixed-mode fracture. The FCM and the EFCM are reformulated to include the shear transfer mechanism on crack surfaces, and the methods are used to study experimental problems.

Chapter 8 focuses on applications in solving real engineering problems, presenting a pseudo-shell model that is developed from the EFCM for crack analysis of tunnel linings. This model is used to calculate the pressure loads acting on various aging waterway tunnels of hydraulic power plants, based on crack analysis. Chapters 9 and 10 present two crack-analysis computer programs for multiple-crack analysis and mixed-mode crack analysis, respectively, including program explanations and illustrative examples.

REFERENCES

ACI Committee 446. (1997). "Finite element analysis of fracture in concrete structures: state-of-the-art." *Report of Subcommittee* 3, ACI, Detroit.

Barenblatt, G. I. (1959). "The formation of equilibrium cracks during brittle fracture: general ideas and hypotheses, axially-symmetric cracks." *J. Appl. Math. Meth.*, 23, 622–636.

Barenblatt, G. I. (1962). "The mathematical theory of equilibrium cracks in brittle fracture." *Advances in Appli. Mech.*, 7, 55–129.

Barpi, F., and Valente, S. (1998). "Size-effects induced bifurcation phenomena during multiple cohesive crack propagation." *J. Solids Struct.*, 35, 1851–1861.

Bazant, Z. P., and Gambarova, P. (1980). "Rough cracks in reinforced concrete." *J. Struc. Div., Proceedings of ASCE*, 106(ST4), 819–842.

Bazant, Z. P., and Ozbolt, J. (1990). "Nonlocal microplane model for fracture, damage and size effect in structures." *J. Eng. Mech.*, 116(11), 2485–2505.

Bazant, Z. P., and Planas, J. (1998). *Fracture and Size Effect: In Concrete and Other Quasibrittle Materials*, CRC Press.

Bazant, Z. P., Tabbara, M. R., Kazemi, M. T., and Pijaudier-Cabot, G. (1990). "Random particle model for fracture of aggregate or fiber composites." *J. Eng. Mech.*, 116(8), 1686–1705.

Bazant, Z. P., and Wahab, A. B. (1980). "Stability of parallel cracks in solids reinforced by bars." *J. Solids Struct.*, 16, 97–105.

Carpinteri, A., and Monetto, I. (1999). "Snap-back analysis of fracture evolution in multi-cracked solids using boundary element method." *Int. J. Fracture*, 98, 225–241.

Cervenka, J. (1994). "Discrete crack modeling in concrete structures." PhD Thesis, University of Colorado, Boulder.

Dugdale, D. S. (1960). "Yielding of steel sheets containing slits." *J. Mech. Phys. Solids*, 8, 100–104.

Gerstle, W. H. and Xie, M. (1992). "FEM modeling of fictitious crack propagation in concrete." *J. Eng. Mech.*, 118(2), 416–434.

Griffith, A. A. (1921). "The phenomena of rupture and flow in solids." *Phil. Trans. Roy. Soc. of London*, A 221, 163–197.

Griffith, A. A. (1924). "The theory of rupture." *Proc. 1st Int. Congress Appl. Mech.*, Biezeno and Burgers eds., pp. 55–63, Waltman.

Hillerborg, A., Modeer, M., and Petersson, P. E. (1976). "Analysis of crack formation and crack growth in concrete by means of fracture mechanics and finite elements." *Cement and Concrete Research*, 6(6), 773–782.

Ingraffea, A. R., Gerstle, W. H., Gergely, P., and Saouma, V. (1984). "Fracture mechanics of bond in reinforced concrete." *J. Struct. Eng.*, 110(4), 871–890.

Ingraffea, A. R., and Saouma, V. (1984). "Numerical modeling of discrete crack propagation in reinforced and plain concrete." *Fracture Mechanics of Concrete: Structural Application and Numerical Calculation*, G. Sih and A. DiTommaso eds., pp. 171–225, Martinus Nijhoff.

Ingraffea, A. R., and Gerstle, W. H. (1985). "Non-Linear fracture models for discrete crack propagation." *Application of Fracture Mechanics to Cementitious Composites*, NATO-ARW, S. P. Shah ed., pp. 247–285, Martinus-Nijhoff.

Murakami, Y. (1993). *Metal Fatigue: Effects of Small Defects and Nonmetallic Inclusions*, pp. 33–53, Yokendo.

Ngo, D., and Scordelis, A. C. (1967). "Finite element analysis of reinforced concrete beams." *ACI Journal, Proceedings*, 64(3), 152–163.

Nilson, A. H. (1967). "Finite element analysis of reinforced concrete." PhD Thesis, Dept. of Civil Engineering, University of California, Berkeley.

Nilson, A. H. (1968). "Nonlinear analysis of reinforced concrete by finite element method." *ACI Journal, Proceedings*, 65(9), 757–766.

Paulay, T., and Loeber, P. J. (1974). "Shear transfer by aggregate interlock." *Shear in Reinforced Concrete, Special Publication* 42, 1, ACI, Detroit, 1–15.

Rashid, Y. R. (1968). "Ultimate strength analysis of reinforced concrete pressure vessels." *Nuclear Engineering and Design*, 7, 334–344.

Reich, R., Plizari, G., Cervenka, J., and Saouma, V. (1993). "Implementation and validation of a nonlinear fracture model in 2D/3D finite element code." *Numerical Models in Fracture Mechanics*, F. H. Wittmann ed., pp. 265–287, Balkema.

Rots, J., and Schellekens, J. (1990). "Interface elements in concrete mechanics." *Computer-Aided Analysis and Design of Concrete Structures*, 2, 909–918, N. Bicanic and H. Mang eds., Pincridge Press.

Shah, S. P., Swartz, S. E., and Ouyang, C. (1995). "Nonlinear fracture mechanics for mode-I quasi-brittle fracture." *Fracture Mechanics of Concrete*, pp. 110–161, Wiley.

Shi, Z. (2004). "Numerical analysis of mixed-mode fracture in concrete using extended fictitious crack model." *J. Struct. Eng.*, 130(11), 1738–1747.

Shi, Z., Ohtsu, M., Suzuki, M., and Hibino, Y. (2001). "Numerical analysis of multiple cracks in concrete using the discrete approach." *J. Struct. Eng.*, 127(9), 1085–1091.

Shi, Z., and Suzuki, M. (2004). "Numerical studies on load-carrying capacities of notched concrete beams subjected to various concentrated loads." *Construction and Building Materials*, 18, 173–180.

Shi, Z., Suzuki, M., and Ohtsu, M. (2004). "Discrete modeling of crack interaction and localization in concrete beams with multiple cracks." *J. Advanced Concrete Technology*, 2(1), 101–111.

Van Mier, J. G. M. (1997). *Fracture Processes of Concrete*, CRC Press.

Xie, M., and Gerstle, W. H. (1995). "Energy-based cohesive crack propagation modeling." *J. Eng. Mech.*, 121(12), 1349–1358.

Linear Elastic and Nonlinear Fracture Mechanics

2

The theory of the fracture mechanics of concrete, which is a branch of nonlinear fracture mechanics (NLFM) with its governing law for crack propagation drawn from the inelastic material behavior exhibited in an extensive fracture process zone (FPZ) ahead of an open crack, is largely developed from the theory of linear elastic fracture mechanics (LEFM). This chapter introduces fundamental concepts from both LEFM and NLFM of concrete that are essential to understanding the subsequent development of the computational theories on multiple-crack analysis. In briefing the basic theories of LEFM, the following textbooks are referenced: *Elementary Engineering Fracture Mechanics* (Broek, 1986), *Fundamentals of Fracture Theories* (Fan, 2003), and *Fracture Mechanics—Fundamentals and Applications* (Anderson, 2005).

2.1 ELASTIC CRACK-TIP FIELDS

This section introduces the elastic theories of the crack-tip stress fields.

2.1.1 Equations of Elasticity and Airy Stress Function

The equations of elasticity and the Airy stress function are derived first in the Cartesian coordinates and then in the polar coordinates.

Cartesian Coordinates

Figure 2.1 illustrates stress components at an arbitrary point (x, y, z) in a deformed body in a Cartesian coordinate system, which are defined as σ_x, σ_y, σ_z, τ_{xy}, τ_{xz}, and τ_{yz}. The corresponding strain components are ε_x, ε_y, ε_z, γ_{xy}, γ_{xz}, and γ_{yz}. For planar problems with the assumption of small displacements, the strain-displacement equations are given by

$$\varepsilon_x = \frac{\partial u_x}{\partial x}; \varepsilon_y = \frac{\partial u_y}{\partial y}; \gamma_{xy} = \frac{\partial u_x}{\partial y} + \frac{\partial u_y}{\partial x} \qquad (2.1)$$

where u_x and u_y are the displacement components in the x and y directions, respectively. To ensure the existence of a unique solution for the displacements, the strain components must satisfy

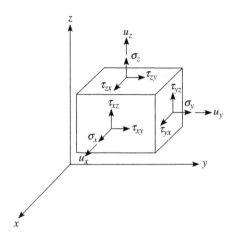

FIGURE 2.1 Stress and displacement components in a Cartesian coordinate system.

the following compatibility equation, which is readily derived from the preceding three equations by suitable differentiation of the strains and interchange of the order of differentiation, as

$$\frac{\partial^2 \varepsilon_x}{\partial y^2} + \frac{\partial^2 \varepsilon_y}{\partial x^2} = \frac{\partial^2 \gamma_{xy}}{\partial x \partial y} \tag{2.2}$$

Assuming the absence of body forces from the problem, the equilibrium equations in the x and y directions are obtained, respectively, as

$$\frac{\partial \sigma_x}{\partial x} + \frac{\partial \tau_{xy}}{\partial y} = 0; \frac{\partial \sigma_y}{\partial y} + \frac{\partial \tau_{xy}}{\partial x} = 0 \tag{2.3}$$

Note that when body forces are present, a solution can first be obtained in the absence of body forces and then modified by superimposing the body forces.

The stress-strain relations for an elastic, isotropic material under the plane stress condition are obtained as

$$\sigma_x = \frac{E}{1-v^2} \left(\varepsilon_x + v\varepsilon_y \right); \sigma_y = \frac{E}{1-v^2} \left(\varepsilon_y + v\varepsilon_x \right); \tau_{xy} = \frac{E}{2(1+v)} \gamma_{xy} \tag{2.4a}$$

$$\sigma_z = \tau_{xz} = \tau_{yz} = 0; \varepsilon_z = \frac{-v}{1-v} \left(\varepsilon_x + \varepsilon_y \right) \tag{2.4b}$$

where E and v are the modulus of elasticity and Poisson's ratio, respectively.

The stress-strain relations under the plane strain condition are given by

$$\sigma_x = \frac{E}{(1+v)(1-2v)} \left[(1-v)\varepsilon_x + v\varepsilon_y \right]; \sigma_y = \frac{E}{(1+v)(1-2v)} \left[(1-v)\varepsilon_y + v\varepsilon_x \right]; \tau_{xy} = \frac{E}{2(1+v)} \gamma_{xy} \tag{2.5a}$$

$$\sigma_z = v \left(\sigma_x + \sigma_y \right); \varepsilon_z = 0; \tau_{xz} = \tau_{yz} = 0 \tag{2.5b}$$

A plane elastic problem is defined completely by a set of eight equations: the three strain-displacement equations (Eqs. 2.1), the two equilibrium equations (Eqs. 2.3), and the three stress-strain relations for either plane stress (Eqs. 2.4a) or plane strain (Eqs. 2.5a). In principle, these eight partial differential equations in eight unknowns (i.e., the three stress components, the three strain components, and the two displacements) could be solved to analytically determine stress-and-strain distributions for any specific boundary-value problem. In reality, however, it is difficult to solve directly these governing equations for a given problem, and the introduction of the following Airy stress function greatly simplifies solutions of planar problems. Assume the stress components can be represented by the Airy stress function, $U(x, y)$, such that

$$\sigma_x = \frac{\partial^2 U}{\partial y^2}; \sigma_y = \frac{\partial^2 U}{\partial x^2}; \tau_{xy} = -\frac{\partial^2 U}{\partial x \partial y} \tag{2.6}$$

Substituting these definitions into Eqs. (2.3), it is immediately known that the equilibrium equations are automatically satisfied. Transforming the compatibility condition of Eq. (2.2) in terms of the Airy stress functions of Eqs. (2.6) with the help of the stress-strain relations (for either plane stress or plane strain), it is found that the Airy stress function must satisfy the following biharmonic equation in the form of

$$\frac{\partial^4 U}{\partial x^4} + 2\frac{\partial^4 U}{\partial x^2 \partial y^2} + \frac{\partial^4 U}{\partial y^4} = 0 \tag{2.7a}$$

or

$$\left(\frac{\partial^2}{\partial x^2} + \frac{\partial^2}{\partial y^2}\right) \cdot \left(\frac{\partial^2}{\partial x^2} + \frac{\partial^2}{\partial y^2}\right) U = 0 \tag{2.7b}$$

where the term in parentheses is the harmonic operator ∇^2. Therefore, for the two-dimensional problem of elasticity under either the plane stress or plane strain condition, the single governing equation in terms of the Airy stress function becomes

$$\nabla^2 \nabla^2 U = 0 \text{ or } \nabla^4 U = 0 \tag{2.7c}$$

As such, the Airy stress function approach integrates all eight of the previous equations into one biharmonic equation, and the solution of a planar problem starts with finding a suitable function $U(x, y)$ that satisfies Eqs. (2.7), subject to the given boundary conditions. Once U is determined, the stresses can be computed from Eqs. (2.6) by differentiation.

Polar Coordinates

Figure 2.2 presents a two-dimensional differential element in equilibrium (assuming unit thickness in the z-direction). Following the conventions in a polar coordinate system, stress, strain and displacement components at any point (r, θ) in a deformed body are defined as $\sigma_r, \sigma_\theta, \tau_{r\theta}, \varepsilon_r, \varepsilon_\theta, \gamma_{r\theta}, u_r$, and u_θ. The strain-displacement equations are obtained as

$$\varepsilon_r = \frac{\partial u_r}{\partial r}; \varepsilon_\theta = \frac{u_r}{r} + \frac{1}{r}\frac{\partial u_\theta}{\partial \theta}; \gamma_{r\theta} = \frac{1}{r}\frac{\partial u_r}{\partial \theta} + \frac{\partial u_\theta}{\partial r} - \frac{u_\theta}{r} \tag{2.8}$$

The compatibility equation in polar form is derived from Eqs. (2.8) and is given by

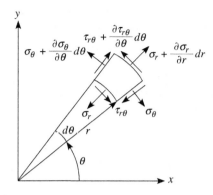

FIGURE 2.2 Stress components on a differential element of unit depth in polar coordinates.

$$\frac{\partial^2 \varepsilon_\theta}{\partial r^2} + \frac{1}{r^2}\frac{\partial^2 \varepsilon_r}{\partial \theta^2} + \frac{2}{r}\frac{\partial \varepsilon_\theta}{\partial r} - \frac{1}{r}\frac{\partial \varepsilon_r}{\partial r} = \frac{1}{r}\frac{\partial^2 \gamma_{r\theta}}{\partial r \partial \theta} + \frac{1}{r^2}\frac{\partial \gamma_{r\theta}}{\partial \theta} \qquad (2.9)$$

Referring to Figure 2.2 and neglecting body forces, the equilibrium equations in the radial and the tangential directions are obtained as

$$\frac{\partial \sigma_r}{\partial r} + \frac{1}{r}\frac{\partial \tau_{r\theta}}{\partial \theta} + \frac{\sigma_r - \sigma_\theta}{r} = 0 \qquad (2.10a)$$

$$\frac{1}{r}\frac{\partial \sigma_\theta}{\partial \theta} + \frac{\partial \tau_{r\theta}}{\partial r} + \frac{2\tau_{r\theta}}{r} = 0 \qquad (2.10b)$$

Since the polar coordinate system is orthogonal, the stress-strain relations can be obtained by substituting r and θ for x and y in Eqs. (2.4a) and Eqs. (2.5a). Thus, the stress-strain relations are given by

$$\sigma_r = \frac{E}{1 - v^2}(\varepsilon_r + v\varepsilon_\theta); \sigma_\theta = \frac{E}{1 - v^2}(\varepsilon_\theta + v\varepsilon_r); \tau_{r\theta} = \frac{E}{2(1 + v)}\gamma_{r\theta} \qquad (2.11a)$$

$$\sigma_z = \tau_{rz} = \tau_{\theta z} = 0; \varepsilon_z = \frac{-v}{1 - v}(\varepsilon_r + \varepsilon_\theta) \qquad (2.11b)$$

for plane stress, and

$$\sigma_r = \frac{E}{(1 + v)(1 - 2v)}[(1 - v)\varepsilon_r + v\varepsilon_\theta]; \sigma_\theta = \frac{E}{(1 + v)(1 - 2v)}[(1 - v)\varepsilon_\theta + v\varepsilon_r]; \tau_{r\theta} = \frac{E}{2(1 + v)}\gamma_{r\theta} \quad (2.12a)$$

$$\sigma_z = v(\sigma_r + \sigma_\theta); \varepsilon_z = 0; \tau_{rz} = \tau_{\theta z} = 0 \qquad (2.12b)$$

for plane strain.

The governing equations in two dimensions have been developed in the polar forms, with a set of eight equations obtained to determine the eight polar unknowns. Next, the Airy stress function, $U(r, \theta)$, is introduced, assuming that

$$\sigma_r = \frac{1}{r}\frac{\partial U}{\partial r} + \frac{1}{r^2}\frac{\partial^2 U}{\partial \theta^2}; \sigma_\theta = \frac{\partial^2 U}{\partial r^2}; \tau_{r\theta} = -\frac{\partial}{\partial r}\left(\frac{1}{r}\frac{\partial U}{\partial \theta}\right) \tag{2.13}$$

It can be verified that with this choice of stress function, the equilibrium equations in polar form, Eqs. (2.10), are satisfied identically. Substituting the stress-strain relations (for either plane stress or plane strain) and the Airy stress functions of Eqs. (2.13) into the compatibility condition of Eq. (2.9), the following Airy biharmonic equation is obtained:

$$\left(\frac{\partial^2}{\partial r^2} + \frac{1}{r}\frac{\partial}{\partial r} + \frac{1}{r^2}\frac{\partial^2}{\partial \theta^2}\right) \cdot \left(\frac{\partial^2}{\partial r^2} + \frac{1}{r}\frac{\partial}{\partial r} + \frac{1}{r^2}\frac{\partial^2}{\partial \theta^2}\right) U = 0 \tag{2.14a}$$

Notice that the term in parentheses is the harmonic operator in polar form. Therefore,

$$\nabla^2 \nabla^2 U = 0 \text{ or } \nabla^4 U = 0 \tag{2.14b}$$

2.1.2 The Williams Solution of Elastic Stress Fields at the Crack Tip

Figure 2.3 illustrates the well-known Williams approach (Williams, 1952, 1957) to an edge-crack problem, whose solution revealed the universal nature of the $r^{-1/2}$ singularity for elastic crack problems and played an important role in the early development of LEFM. Williams first solved the wedge problem with an arbitrary apex angle of 2α, as shown in Figure 2.3a and later extended the solution to the case of a sharp crack by letting $\alpha = \pm\pi$, as shown in Figure 2.3b.

Instead of adopting a closed-form function, Williams assumed an Airy stress function in polar coordinates in the form of the near-tip asymptotic series expansion, as

$$U(r, \theta) = \sum_n r^{n+1} F_n(\theta) \tag{2.15}$$

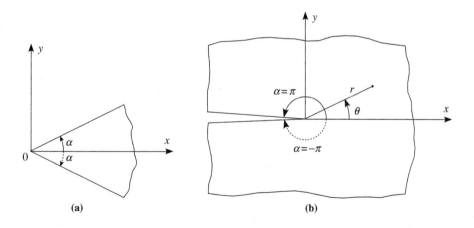

(a)   **(b)**

FIGURE 2.3 Williams approach to edge-crack problems: (a) a wedge of apex angle 2α with traction-free flanks and (b) special case of a sharp crack with $\alpha = \pm\pi$.

Based on Eqs. (2.13), the stresses in polar coordinates are obtained as

$$\sigma_r = \sum_n r^{n-1}\left[(n+1)F_n(\theta) + F_n''(\theta)\right]; \sigma_\theta = \sum_n r^{n-1}[n(n+1)F_n(\theta)]; \tau_{r\theta} = \sum_n r^{n-1}\left[-nF_n'(\theta)\right] \quad (2.16)$$

Substituting the assumed Airy stress function of Eq. (2.15) into the Airy bi-harmonic equation, Eqs. (2.14), and requiring that $F_n(\theta)$ be independent of the coordinate, r, leads to

$$\frac{d^4F_n(\theta)}{d\theta^4} + \left[(n+1)^2 + (n-1)^2\right]\frac{d^2F_n(\theta)}{d\theta^2} + \left[(n+1)^2(n-1)^2\right]F_n(\theta) = 0 \quad (2.17a)$$

or in compact form as

$$\left[\frac{d^2}{d\theta^2} + (n+1)^2\right] \cdot \left[\frac{d^2}{d\theta^2} + (n-1)^2\right]F_n(\theta) = 0 \quad (2.17b)$$

Judging from Eqs. (2.17), $F_n(\theta)$ is assumed to take the following form:

$$F_n(\theta) = A_n \sin(n+1)\theta + B_n \cos(n+1)\theta + C_n \sin(n-1)\theta + D_n \cos(n-1)\theta \quad (2.18)$$

where the values of n, the constants of A_n, B_n, C_n, and D_n, are to be determined. The stress-free condition on the crack surfaces requires that $\sigma_\theta(\pi) = \sigma_\theta(-\pi) = \tau_{r\theta}(\pi) = \tau_{r\theta}(-\pi) = 0$, which implies the following boundary conditions:

$$F_n(\pi) = F_n(-\pi) = 0; F_n'(\pi) = F_n'(-\pi) = 0 \quad (2.19)$$

Substituting Eq. (2.18) into Eqs. (2.19) leads to a system of simultaneous equations for the unknown constants. To obtain meaningful solutions from the resulting homogeneous equations, the determinants of the coefficient matrices must be equal to zero. This requires that

$$\sin 2n\pi = 0 \text{ where } n = 0, \pm\frac{1}{2}, \pm1, \pm\frac{3}{2}, \pm2, \ldots \quad (2.20)$$

The values of n need to be further examined to eliminate invalid solutions. The strain energy accumulated in an element of unit thickness is known as

$$dW \propto \sigma_{ij}^2 r dr d\theta \propto r^{2n-2} r dr d\theta \quad (2.21a)$$

Let W represent the total strain energy of a finite body encircled by the radius R. Therefore,

$$W \propto \int_0^R r^{2n-1}dr \quad (2.21b)$$

The finiteness of the total strain energy W is guaranteed only when n satisfies the following condition:

$$2n - 1 > -1 \quad (2.21c)$$

Eq. (2.21c) defines the validity range of n as

$$n = \frac{1}{2}, 1, \frac{3}{2}, 2, \ldots \quad (2.21d)$$

Based on Eq. (2.18), Eqs. (2.19) and Eq. (2.21d), it is found that the four constants of A_n, B_n, C_n, and D_n are interconnected through the following relations:

$$B_n = -\frac{n-2}{n+2}D_n \text{ and } A_n = -C_n \text{ for } n = 1,3,5,\ldots \tag{2.22a}$$

$$B_n = -D_n \text{ and } A_n = -\frac{n-2}{n+2}C_n \text{ for } n = 2,4,6,\ldots \tag{2.22b}$$

From these relations and the valid values of n, the general solution for the Airy stress function is obtained as the series

$$
\begin{aligned}
U(r,\theta) = \sum_{n=1,3,\ldots} r^{1+\frac{n}{2}} &\left[D_n\left(\cos\frac{n-2}{2}\theta - \frac{n-2}{n+2}\cos\frac{n+2}{2}\theta\right) + C_n\left(\sin\frac{n-2}{2}\theta - \sin\frac{n+2}{2}\theta\right) \right] \\
+ \sum_{n=2,4,\ldots} r^{1+\frac{n}{2}} &\left[D_n\left(\cos\frac{n-2}{2}\theta - \cos\frac{n+2}{2}\theta\right) + C_n\left(\sin\frac{n-2}{2}\theta - \frac{n-2}{n+2}\sin\frac{n+2}{2}\theta\right) \right]
\end{aligned}
\tag{2.23}
$$

Williams split the obtained Airy stress function into its even and odd parts with respect to the crack plane ($\theta = 0$), and he derived the dominant terms of the crack-tip stress fields as

$$\sigma_r = \frac{1}{4\sqrt{r}}\left(5\cos\frac{\theta}{2} - \cos\frac{3\theta}{2}\right)D_n; \quad \sigma_\theta = \frac{1}{4\sqrt{r}}\left(3\cos\frac{\theta}{2} + \cos\frac{3\theta}{2}\right)D_n;$$

$$\tau_{r\theta} = \frac{1}{4\sqrt{r}}\left(\sin\frac{\theta}{2} + \sin\frac{3\theta}{2}\right)D_n \tag{2.24}$$

for symmetric loading (opening mode), and

$$\sigma_r = \frac{1}{4\sqrt{r}}\left(-5\sin\frac{\theta}{2} + 3\sin\frac{3\theta}{2}\right)C_n; \quad \sigma_\theta = \frac{1}{4\sqrt{r}}\left(-3\sin\frac{\theta}{2} - 3\sin\frac{3\theta}{2}\right)C_n$$

$$\tau_{r\theta} = \frac{1}{4\sqrt{r}}\left(\cos\frac{\theta}{2} + 3\cos\frac{3\theta}{2}\right)C_n \tag{2.25}$$

for antisymmetric loading (in plane shear). It should be noted that the preceding equations are derived without assuming any specific configuration for the problem, and therefore the inverse-square-root singularity exhibited by the crack-tip stresses reveals a universal property for cracks in isotropic elastic materials. Obviously, the constants D_n and C_n depend on the loading condition as well as the specific configuration of the problem, including the size of the crack.

It is known that under symmetric loading, such as pure bending or pure tension, an opening-mode crack (Mode I) develops, and under antisymmetric loading, such as in-plane shear, a shear-mode crack (Mode II) propagates. When a crack is subjected to both symmetric and anti-symmetric loading as in most practical cases (as the Williams solution indicates), the resulting near-tip stress fields are given by the superposition of the stresses from Eqs. (2.24) and (2.25), and the crack is called a mixed-mode crack (Mode I + Mode II).

2.1.3 **The Complex Stress Function Approach to Elastic Stress Fields at the Crack Tip**

In this section, the complex stress function approach is introduced, and the method is used to obtain the elastic stress fields at the crack tip.

The Complex Stress Function for Plane Elastic Problems

The use of complex variables in solving plane elastic problems has a solid theoretical foundation. The following formulation is based on the work of Muskhelishvili (1953). A complex function $f(z)$ is defined by

$$f(z) = P(x, y) + iQ(x, y) \qquad (2.26)$$

where $z = x + iy$; $P(x, y)$ is the real part of $f(z)$—that is, $P(x, y) = \text{Re} f(z)$; and $Q(x, y)$ is the imaginary part of $f(z)$—that is, $Q(x, y) = \text{Im} f(z)$. The function $f(z)$ is analytic if,

$$\frac{\partial P}{\partial x} = \frac{\partial Q}{\partial y}; \frac{\partial P}{\partial y} = -\frac{\partial Q}{\partial x} \qquad (2.27)$$

These equations are called the Cauchy-Riemann conditions and are derivable based on the assumption that $f(z)$ is single-valued and possesses a unique derivative at every point in a given domain. Based on Eqs. (2.27), it is immediately verifiable that the real and imaginary parts of any analytical function are harmonic functions—that is,

$$\nabla^2 P = \left(\frac{\partial^2}{\partial x^2} + \frac{\partial^2}{\partial y^2}\right) P = 0; \nabla^2 Q = \left(\frac{\partial^2}{\partial x^2} + \frac{\partial^2}{\partial y^2}\right) Q = 0 \qquad (2.28)$$

Using the complex variable z and its conjugate, $\bar{z} = x - iy$ an Airy stress function $U(x, y)$ can be expressed in terms of complex variables, $U(x, y) = U(z, \bar{z})$. Based on the following facts,

$$\frac{\partial z}{\partial x} = 1; \frac{\partial z}{\partial y} = i; \frac{\partial \bar{z}}{\partial x} = 1; \frac{\partial \bar{z}}{\partial y} = -i \qquad (2.29)$$

it can be derived that

$$\frac{\partial^2 U}{\partial x^2} = \left(\frac{\partial}{\partial z} + \frac{\partial}{\partial \bar{z}}\right)^2 U; \frac{\partial^2 U}{\partial y^2} = -\left(\frac{\partial}{\partial z} - \frac{\partial}{\partial \bar{z}}\right)^2 U; \nabla^2 U = \frac{\partial^2 U}{\partial x^2} + \frac{\partial^2 U}{\partial y^2} = 4\frac{\partial^2 U}{\partial z \partial \bar{z}} \qquad (2.30)$$

This leads to

$$\nabla^2 \nabla^2 U = 16\frac{\partial^4 U}{\partial z^2 \partial \bar{z}^2} = 0, \text{i.e.,} \frac{\partial^4 U}{\partial z^2 \partial \bar{z}^2} = 0 \qquad (2.31)$$

Integrating Eq. (2.31) with respect to z and $\bar{z}$ twice, respectively, leads to

$$U = f_1(z) + \bar{z} f_2(z) + f_3(\bar{z}) + z f_4(\bar{z}) \qquad (2.32)$$

where f_1, f_2, f_3, and f_4 are arbitrary complex functions. Since U is a real function, f_1 and f_3, and f_2 and f_4 must be conjugate pairs having the following relations:

$$f_3(\bar{z}) = \overline{f_1(z)}; f_4(\bar{z}) = \overline{f_2(z)} \qquad (2.33)$$

Eq. (2.32) is then rewritten as

$$U = f_1(z) + \bar{z}f_2(z) + \overline{f_1(z)} + z\overline{f_2(z)}$$ (2.34)

By introducing the transformation of

$$f_1(z) = \frac{1}{2}\chi(z); \, f_2(z) = \frac{1}{2}\phi(z)$$ (2.35)

the Airy stress function U is expressed in terms of two analytic functions of ϕ and χ, as

$$U = \frac{1}{2}\left[\bar{z}\phi(z) + z\overline{\phi(z)} + \chi(z) + \overline{\chi(z)}\right]$$ (2.36)

or in its equivalent form of

$$U = \mathrm{Re}[\bar{z}\phi(z) + \chi(z)]$$ (2.37)

Substituting Eq. (2.36) into Eqs. (2.6), the stresses are derived as

$$\sigma_x + \sigma_y = 2\left[\phi'(z) + \overline{\phi'(z)}\right]$$ (2.38a)

$$\sigma_y - \sigma_x + 2i\tau_{xy} = 2[\bar{z}\phi''(z) + \chi''(z)] = 2[\bar{z}\phi''(z) + \psi'(z)]$$ (2.38b)

where a function with a prime represents its derivative with respect to z and $\psi(z) = \chi'(z)$ is assumed. The stress components can be obtained by separating the real and imaginary parts of Eqs. (2.38). The displacement components are derived from the stress-strain relations of Eqs. (2.4) and (2.5), as

$$\frac{E}{1+v}(u_x + iu_y) = \frac{3-v}{1+v}\phi(z) - z\overline{\phi'(z)} - \overline{\psi(z)}$$ (2.39)

for plane stress, and

$$\frac{E}{1+v}(u_x + iu_y) = (3 - 4v)\phi(z) - z\overline{\phi'(z)} - \overline{\psi(z)}$$ (2.40)

for plane strain.

The Central Crack Problem

A crack in an infinite plate under tension is illustrated in Figure 2.4. The following complex functions, ϕ and ψ, are obtained through the conformal transformation of the solutions of an equivalent problem concerning a unit circle in an auxiliary plane, as

$$\phi(z) = \frac{\sigma}{4}\left(2\sqrt{z^2 - a^2} - z\right); \, \psi(z) = \frac{\sigma}{2}\left(z - \frac{a^2}{\sqrt{z^2 - a^2}}\right)$$ (2.41)

It can be verified that ϕ and ψ satisfy Eqs. (2.27), as well as the boundary conditions. The stress components are obtained by substituting Eqs. (2.41) into Eqs. (2.38) as

$$\sigma_x + \sigma_y = \sigma\left(2\mathrm{Re}\frac{z}{\sqrt{z^2 - a^2}} - 1\right); \, \sigma_y - \sigma_x + 2i\tau_{xy} = \sigma\left(\frac{2ia^2y}{\sqrt{(z^2 - a^2)^3}} + 1\right)$$ (2.42)

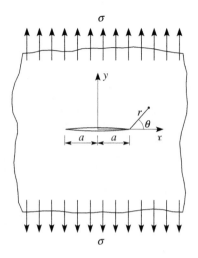

FIGURE 2.4 An infinite plate with a central crack subjected to tension.

To obtain the near-tip stress fields, the following transformations between Cartesian coordinates and polar coordinates, as shown in Figure 2.4, are introduced for convenience:

$$x = a + r\cos\theta; \ y = r\sin\theta; \ z = a + r\cos\theta + ir\sin\theta \tag{2.43}$$

Substituting Eqs. (2.43) into Eqs. (2.42) and omitting higher order terms of r lead to

$$\sigma_x + \sigma_y = \sigma\sqrt{\frac{2a}{r}}\cos\frac{\theta}{2}$$

$$\sigma_y - \sigma_x + 2i\tau_{xy} = \sigma\sqrt{\frac{2a}{r}}\sin\frac{\theta}{2}\cos\frac{\theta}{2}\left(\sin\frac{3\theta}{2} + i\cos\frac{3\theta}{2}\right) \tag{2.44}$$

After separating the real and imaginary parts from Eqs. (2.44), the near-tip stress fields for a central crack subjected to tension forces are obtained as

$$\sigma_x = \sigma\frac{\sqrt{a}}{\sqrt{2r}}\cos\frac{\theta}{2}\left(1 - \sin\frac{\theta}{2}\sin\frac{3\theta}{2}\right); \ \sigma_y = \sigma\frac{\sqrt{a}}{\sqrt{2r}}\cos\frac{\theta}{2}\left(1 + \sin\frac{\theta}{2}\sin\frac{3\theta}{2}\right)$$

$$\tau_{xy} = \sigma\frac{\sqrt{a}}{\sqrt{2r}}\sin\frac{\theta}{2}\cos\frac{\theta}{2}\cos\frac{3\theta}{2} \tag{2.45}$$

Similarly, for a crack in an infinite plate subjected to shear, as shown in Figure 2.5, the near-tip stress fields are given by

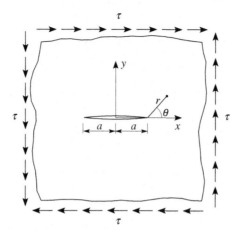

FIGURE 2.5 An infinite plate with a central crack subjected to shear.

$$\sigma_x = -\tau \frac{\sqrt{a}}{\sqrt{2r}} \sin\frac{\theta}{2} \left(2 + \cos\frac{\theta}{2}\cos\frac{3\theta}{2} \right); \; \sigma_y = \tau \frac{\sqrt{a}}{\sqrt{2r}} \sin\frac{\theta}{2}\cos\frac{\theta}{2}\cos\frac{3\theta}{2}$$

$$\tau_{xy} = \tau \frac{\sqrt{a}}{\sqrt{2r}} \cos\frac{\theta}{2} \left(1 - \sin\frac{\theta}{2}\sin\frac{3\theta}{2} \right)$$

(2.46)

It should be noted that the crack-tip stresses defined by Eqs. (2.45) and (2.46) in Cartesian coordinates are equivalent to those defined by Eqs. (2.24) and (2.25) in polar coordinates, even though the Williams edge-crack problem and the present central-crack problem with an internal crack in an elastic body represent two different types of fracture problems with distinct geometric differences. Consequently, it is concluded that, under the assumption of elasticity, the crack-tip stresses have an invariant form of distribution with an inverse-square-root singularity at the tip of the crack.

As clearly suggested by the Williams general solution, for each specific problem of interest, the remaining task is to determine the unknown constant in Eqs. (2.24) or Eqs. (2.25) so as to satisfy the given load and boundary conditions. This single constant defines the amplitude of the stresses in the vicinity of the crack tip, with which the crack-tip stress fields are uniquely determined. The significance of this constant in characterizing the crack-tip fields in an elastic body was first recognized by Irwin (1957), who called it the stress intensity factor, *K*, which has become an important fracture mechanics parameter and a fundamental concept in the theory of LEFM.

2.2 STRESS INTENSITY FACTOR AND *K*-CONTROLLED CRACK-TIP FIELDS

A crack may be subjected to three different types of loading that cause displacements of the crack surfaces, as Figure 2.6 illustrates. Mode I loading, where the load is applied normal to the crack plane, tends to open the crack. Mode II loading refers to in-plane shear and causes

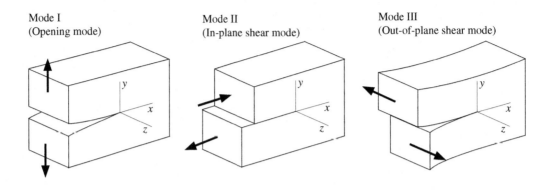

FIGURE 2.6 Three independent modes of deformation at the crack tip.

the two crack surfaces to slide against each other. Mode III loading, where out-of-plane shear is applied, tends to tear the two crack surfaces apart. This last mode of deformation, called out-of-plane shear mode, does not occur in the plane elastic problem. The stress intensity factors for the three types of loading are denoted as K_I, K_{II}, and K_{III}, respectively. The crack-tip stresses for each mode of loading can now be rewritten in terms of the corresponding stress intensity factor, as

Mode I

$$\sigma_x = \frac{K_I}{\sqrt{2\pi r}} \cos\frac{\theta}{2}\left(1 - \sin\frac{\theta}{2}\sin\frac{3\theta}{2}\right)$$

$$\sigma_y = \frac{K_I}{\sqrt{2\pi r}} \cos\frac{\theta}{2}\left(1 + \sin\frac{\theta}{2}\sin\frac{3\theta}{2}\right) \qquad (2.47)$$

$$\tau_{xy} = \frac{K_I}{\sqrt{2\pi r}} \sin\frac{\theta}{2}\cos\frac{\theta}{2}\cos\frac{3\theta}{2}$$

Mode II

$$\sigma_x = -\frac{K_{II}}{\sqrt{2\pi r}} \sin\frac{\theta}{2}\left(2 + \cos\frac{\theta}{2}\cos\frac{3\theta}{2}\right)$$

$$\sigma_y = \frac{K_{II}}{\sqrt{2\pi r}} \sin\frac{\theta}{2}\cos\frac{\theta}{2}\cos\frac{3\theta}{2} \qquad (2.48)$$

$$\tau_{xy} = \frac{K_{II}}{\sqrt{2\pi r}} \cos\frac{\theta}{2}\left(1 - \sin\frac{\theta}{2}\sin\frac{3\theta}{2}\right)$$

where $\tau_{xz} = \tau_{yz} = 0$ for plane stress and plane strain; and $\sigma_z = 0$ for plane stress and $\sigma_z = v(\sigma_x + \sigma_y)$ for plane strain.

Mode III

$$\tau_{xz} = -\frac{K_{III}}{\sqrt{2\pi r}}\sin\frac{\theta}{2}$$

$$\tau_{yz} = \frac{K_{III}}{\sqrt{2\pi r}}\cos\frac{\theta}{2}$$

(2.49)

where $\sigma_x = \sigma_y = \sigma_z = \tau_{xy} = 0$.

Similarly, the crack-tip displacement fields are obtained as

Mode I

$$u_x = \frac{K_I}{2G}\sqrt{\frac{r}{2\pi}}\cos\frac{\theta}{2}\left(\kappa - 1 + 2\sin^2\frac{\theta}{2}\right)$$

$$u_y = \frac{K_I}{2G}\sqrt{\frac{r}{2\pi}}\sin\frac{\theta}{2}\left(\kappa + 1 - 2\cos^2\frac{\theta}{2}\right)$$

(2.50)

Mode II

$$u_x = \frac{K_{II}}{2G}\sqrt{\frac{r}{2\pi}}\sin\frac{\theta}{2}\left(\kappa + 1 + 2\cos^2\frac{\theta}{2}\right)$$

$$u_y = -\frac{K_{II}}{2G}\sqrt{\frac{r}{2\pi}}\cos\frac{\theta}{2}\left(\kappa - 1 - 2\sin^2\frac{\theta}{2}\right)$$

(2.51)

Mode III

$$u_z = \frac{2K_{III}}{G}\sqrt{\frac{r}{2\pi}}\sin\frac{\theta}{2}$$

(2.52)

where G is the shear modulus; $\kappa = 3 - 4v$ for plane strain, and $\kappa = (3 - v)/(1 + v)$ for plane stress; and $E = 2(1 + v)G$.

By comparing Eqs. (2.47) and (2.48) with Eqs. (2.45) and (2.46), which are the exact solutions for an infinite plate with a central crack as $r \to 0$, the stress intensity factors for this particular problem are found to be

$$K_I = \sigma\sqrt{\pi a}; \ K_{II} = \tau\sqrt{\pi a}$$

(2.53)

As seen, the stress intensity factor is a function of the applied load and the size of the crack; when a finite body is involved, it is also a function of the geometric configuration of the problem. Obviously, the proportionality of K to the applied load reflects the linear nature of the theory of elasticity. In general, the stress intensity factors can be defined as

$$K_I = \lim_{r\to 0}\sqrt{2\pi r}\sigma_y(r, \theta = 0) \ K_{II} = \lim_{r\to 0}\sqrt{2\pi r}\tau_{xy}(r, \theta = 0)$$

$$K_{III} = \lim_{r\to 0}\sqrt{2\pi r}\tau_{yz}(r, \theta = 0)$$

(2.54)

According to Eqs. (2.54), the stress intensity factor stipulates the magnitude of the stress singularity at the crack tip. The larger the applied load and the size of the crack are (resulting in a larger K), the greater the rate of stress increase in the singularity-dominated zone becomes, provided that the crack remains stationary. As seen in Eqs. (2.47) to (2.52), the stress intensity factor completely defines the crack-tip fields (or the stress redistribution due to the occurrence of the crack); if K is known, it is possible to solve for all the components of stress, strain, and displacement as functions of r and θ.

As a function of the applied load and the size of the crack as well as the geometry of the problem, the stress intensity factor represents both the strength of the crack-tip singularity and the mode of deformation, and therefore it is regarded as an important fracture mechanics parameter in LEFM that provides a link between the crack-contained local material behavior and the global structural response. Based on the principle of linear superposition, the crack-tip stress fields of a mixed-mode crack can be written as

$$\sigma_{ij}(r,\theta) = \frac{1}{\sqrt{2\pi r}} \left\{ K_I f_{ij}^I(\theta) + K_{II} f_{ij}^{II}(\theta) + K_{III} f_{ij}^{III}(\theta) \right\} \tag{2.55}$$

where the functions of $f_{ij}^I(\theta)$, $f_{ij}^{II}(\theta)$, and $f_{ij}^{III}(\theta)$ are defined by Eqs. (2.47) to (2.49).

Although the K-controlled crack-tip fields ($r << a$) have been obtained using the theory of elasticity, these solutions do not reflect the physical reality of the material deformation at the tip of a crack in engineering materials because no real material can sustain infinite stresses, as the inverse-square-root singularity in the stress solutions predicts. It is known that the singularity in the elastic solutions is the result of two mathematical simplifications: the assumption that the configuration of a real crack (whose crack-tip radius is finite) is a sharp crack (whose crack-tip radius is zero) and the assumption that the inelastic material behavior at the crack tip is elastic. To remove the crack-tip singularity, it is necessary to take into account inelastic material behavior, such as plasticity in metals and fracture process zone (FPZ) in concrete, by introducing the inelastic-zone concept to the crack tip. Despite the inaccuracy in describing the actual material behavior in the close vicinity of the crack tip, the elastic solutions of Eqs. (2.47) to (2.52) are of great importance in LEFM. The singularity-based solution of the stress distribution at the tip of a sharp crack in a linear elastic material reveals the nature of fracture in its ultimate form—that is, causing unbounded stresses at the crack tip.

With the help of these elastic solutions, Irwin (1957) envisioned the physical significance of the constants contained in these solutions and coined the term *stress intensity factor* to characterize the singularity at the crack tip, which has turned out to be one of the most important concepts in fracture mechanics. Furthermore, these elastic solutions may still be valid outside the inelastic zone at the crack tip if the size of the zone is sufficiently small as compared to the length dimensions of the cracked structure. Needless to say, these solutions also provide an effective means for calculating the stress intensity factors, as suggested by Eqs. (2.54).

Outside the K-controlled crack-tip fields, the general solutions for stress or displacement components include higher-order terms of r. For example, the solution of stress for any type of loading (K) is given by

$$\sigma_{ij}(r,\theta) = \frac{K}{\sqrt{2\pi r}} f_{ij}(\theta) + \sum_{m=0}^{\infty} A_m r^{\frac{m}{2}} h_{ij}^{(m)}(\theta) \tag{2.56}$$

where A_m is the amplitude for the mth term and $h_{ij}^{(m)}$ is a dimensionless function of θ. Note that the higher-order terms depend on the geometry of the problem. As $r \to 0$, the leading term approaches infinity, while the other terms remain finite or approach zero.

Analytical and numerical methods for calculating stress intensity factors have been extensively discussed in many textbooks of fracture mechanics, and handbooks on K of different geometries and loading conditions are also available. This book focuses on the derivation and physical interpretation of the stress intensity factors and the K-controlled crack-tip fields because of their theoretical importance for the subsequent development of nonlinear fracture mechanics of concrete; for detailed coverage on stress intensity factors, refer to the textbooks cited previously.

2.3 THE ENERGY PRINCIPLES

With the crack-tip fields now being completely defined in terms of the stress intensity factor, an energy description of the fracture process needs to be explained using Griffith's fracture theory (1921, 1924), whose innovative concept of introducing fracture energy into the study of cracked materials laid a solid foundation for the later development of fracture mechanics. It will be shown that the fracture energy required for creating a unit surface of an open crack—that is, the energy release rate—has a close relationship with the stress intensity factor.

2.3.1 The Griffith Fracture Theory

Griffith derived his fracture criterion by considering the problem shown in Figure 2.4, where an infinite plate of unit thickness that contains a crack of length $2a$ is subjected to uniform tensile stress σ. He first obtained the decrease of strain energy in the plate due to the introduction of the crack based on the stress analysis of Inglis (1913), as

$$U_a = -\frac{\pi a^2 \sigma^2}{E} \ (\text{plane stress}) \tag{2.57a}$$

$$U_a = -\frac{1 - v^2}{E} \pi a^2 \sigma^2 \ (\text{plane strain}) \tag{2.57b}$$

where E is the modulus of elasticity, and v is Poisson's ratio. Griffith then introduced "the fundamental conception of the new theory"—that is, the surface energy, which is the fracture energy required to form the open crack—as

$$U_s = 4a\gamma_s \tag{2.58}$$

where γ_s is the surface energy per unit area. Notice that in Eq. (2.58) the crack surface is $4a$ because a crack has two surfaces. According to the Griffith energy criterion, crack growth is a natural course of energy transfer between the strain energy of an elastic body and the fracture energy required for creating a new crack surface. This requires the following condition to be satisfied:

$$\frac{d}{dA}(U_a + U_s) = \frac{1}{2}\frac{d}{da}(U_a + U_s) = 0 \tag{2.59}$$

Therefore, the critical load for crack propagation is obtained as

$$\sigma_c = \sqrt{\frac{2E\gamma_s}{\pi a}} \qquad (2.60a)$$

in the case of plane stress, and

$$\sigma_c = \sqrt{\frac{2E\gamma_s}{(1-v^2)\pi a}} \qquad (2.60b)$$

in the case of plane strain. Equation (2.60) is referred to as the Griffith fracture criterion, which is valid for ideally brittle materials.

2.3.2 **The Energy Release Rate** *G*

The Griffith energy concept is of a global character: For a crack to propagate, the energy stored in the system must be sufficient to overcome the fracture energy of the material. Based on the Griffith fracture theory, Irwin generalized the concept by defining an energy release rate *G*, which is a measure of the energy available for a unit extension of the crack and represents "the force tending to cause crack extension" (Irwin, 1957).

Figure 2.7 shows a cracked plate under a given load *P*, which causes the crack to extend by an amount of *dA*. Since the load is fixed at *P*, the system is under load control. Obviously, the fracture energy dissipated in propagating the crack originates from the work done by the external force *PdΔ* and the released elastic strain energy *–dU* in the system—that is,

$$G = P\frac{d\Delta}{dA} - \frac{dU}{dA} \qquad (2.61)$$

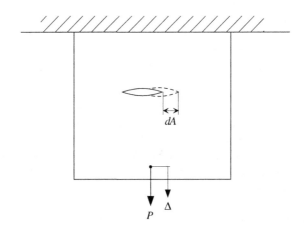

FIGURE 2.7 Cracked plate under a load *P*.

The strain energy of the elastic system is given by

$$U = \int_0^\Delta P d\Delta = \frac{P\Delta}{2} = \frac{CP^2}{2} \tag{2.62}$$

where the relation $\Delta = CP$ is employed; C is the compliance of the system, which is the inverse of the plate stiffness. Notice that

$$\frac{d}{dA} = \frac{\partial}{\partial A} + \frac{dP}{dA}\frac{\partial}{\partial P} \tag{2.63}$$

After substituting Eq. (2.62) and $\Delta = CP$ into Eq. (2.61), the preceding derivative operations are carried out and the energy release rate is derived as

$$G = \frac{P^2}{2}\frac{\partial C}{\partial A} \tag{2.64}$$

It can be proved that Eq. (2.64) can be obtained under the condition of displacement control as well. Theoretically, if the derivative of the compliance of a test specimen with respect to the crack length can be derived experimentally or numerically by an FE solution, then the energy release rate can be calculated from Eq. (2.64).

2.3.3 Relationship between *K* and *G*

Two fracture parameters have been introduced so far: the stress intensity factor K and the energy release rate G. While the first parameter uniquely defines the near-tip stress and displacement fields, the second one represents the crack driving force to open that crack. Obviously, a relationship exists between the two. The following deduction is based on Irwin's crack closure analysis (1957). Figure 2.8 shows an infinite plate of unit thickness with fixed ends that contains a crack

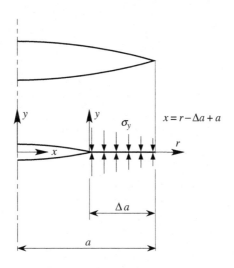

FIGURE 2.8 Crack closure analysis.

of size a. Since the problem is a mode-I fracture, the terms K and G are denoted as K_I and G_I, respectively. Suppose that compressive forces are applied to the crack edge over an infinitesimal distance Δa to close the crack. Since the work required to close the crack is equal to the fracture energy needed to extend the crack, the energy release rate can be calculated from

$$G_I = \lim_{\Delta a \to 0} \frac{\Delta W}{\Delta a} = \lim_{\Delta a \to 0} \frac{2}{\Delta a} \int_0^{\Delta a} \frac{1}{2} \sigma_y u_y dr \qquad (2.65)$$

where ΔW is the required work, and u_y is the displacement in the y direction. Notice that the factor 2 arises because the crack has two surfaces and that the incremental work at r is calculated based on the fact that the stresses increase from zero.

For the infinite plate under tension, the value of u_y is given as

$$u_y = \frac{2\sigma}{E} \sqrt{a^2 - x^2} = \frac{2K_I}{E} \sqrt{\frac{a^2 - x^2}{\pi a}} \qquad (2.66)$$

Substituting $x = r - \Delta a + a$ into Eq. (2.66) and neglecting second-order terms lead to

$$u_y = \frac{2K_I}{E\sqrt{\pi}} \sqrt{2\Delta a - 2r + \frac{2r\Delta a}{a} - \frac{r^2}{a}} \cong \frac{2K_I}{E\sqrt{\pi}} \sqrt{2(\Delta a - r)} \qquad (2.67)$$

The closing force is given by

$$\sigma_y = \frac{K_I}{\sqrt{2\pi r}} \qquad (2.68)$$

By using Eqs. (2.67) and (2.68), Eq. (2.65) becomes

$$G_I = \lim_{\Delta a \to 0} \frac{2K_I^2}{E\pi\Delta a} \int_0^{\Delta a} \sqrt{\frac{\Delta a - r}{r}} dr \qquad (2.69)$$

which leads to

$$G_I = \frac{K_I^2}{E} \text{ (plane stress)} \qquad (2.70a)$$

and to

$$G_I = \left(1 - v^2\right) \frac{K_I^2}{E} \text{ (plane strain)} \qquad (2.70b)$$

The obtained relation between G and K clearly reveals that the stress intensity factor has an energy basis. Similarly, it can be shown that for mode II:

$$G_{II} = \frac{K_{II}^2}{E} \text{ (plane stress)} \qquad (2.71a)$$

$$G_{II} = \left(1 - v^2\right) \frac{K_{II}^2}{E} \text{ (plane strain)} \qquad (2.71b)$$

and for mode III:

$$G_{III} = \frac{1}{1-v}\frac{K_{III}^2}{E} \text{ (plane stress)} \tag{2.72a}$$

$$G_{III} = (1+v)\frac{K_{III}^2}{E} \text{ (plane strain)} \tag{2.72b}$$

The total energy release rate in combined crack modes can be obtained by superposition:

$$G = G_I + G_{II} + G_{III} = \frac{1}{E}\left(K_I^2 + K_{II}^2 + \frac{1}{1-v}K_{III}^2\right) \text{(plane stress)} \tag{2.73a}$$

$$G = G_I + G_{II} + G_{III} = \frac{1-v^2}{E}\left(K_I^2 + K_{II}^2 + \frac{1}{1-v}K_{III}^2\right) \text{(plane strain)} \tag{2.73b}$$

Since the K-controlled near-tip fields are uniquely defined for linear elastic materials, it can be proved that Eqs. (2.70) to (2.72) are valid for all geometries of specimens and types of cracks. In fact, G may even be considered as a physical parameter defined at the tip of a crack using the near-tip stress and displacement fields, as shown in Eq. (2.65). Hence, its relations with K must be invariant, as the near-tip fields are uniquely established, regardless of any specific changes that may occur to the problem. For plates of finite size, the influence of the geometric shape of the problem is reflected in the stress intensity factor K. Therefore, it is possible to obtain G for a problem of finite size by using the relationships between G and K.

2.3.4 The Criterion for Crack Propagation

Since the stress at the tip of a crack in a K-controlled stress field is unbounded, a criterion for crack propagation cannot be stipulated in terms of the tip stress but must be described by using either of the two fracture parameters introduced so far—the energy release rate or the stress intensity factor—which are equivalent based on Eqs. (2.70) to (2.72). For ideally brittle materials, the Griffith criterion in Eq. (2.60a) can be rewritten in terms of the critical energy release rate, G_{Ic}, as

$$\sigma_c = \sqrt{\frac{EG_{Ic}}{\pi a}} \tag{2.74}$$

where

$$G_{Ic} = \frac{dU_s}{dA} = 2\gamma_s \tag{2.75}$$

Substituting Eq. (2.70a) into Eq. (2.74), the critical load is then given in terms of the critical stress intensity factor, K_{Ic}, as

$$\sigma_c = \frac{K_{Ic}}{\sqrt{\pi a}} \tag{2.76}$$

or

$$K_I = K_{Ic} \tag{2.77}$$

Equation (2.76) states that a crack propagates whenever K_I is equal to the threshold value, K_{Ic}, which is a measure of fracture toughness and is regarded as a material fracture property based on LEFM. In principle, the criterion for mode-II or mode-III shear-type fracture can also be expressed as

$$K_{II} = K_{IIc} \tag{2.78}$$

$$K_{III} = K_{IIIc} \tag{2.79}$$

For real engineering materials, stresses cannot increase unboundedly, and yielding takes place in metallic materials to form a plastic zone ahead of an open crack. Therefore, the K-controlled stress states in Eqs. (2.47) to (2.52) do not describe the stress distribution inside the plastic zone. This raises a serious question: Is the K-criterion still a valid criterion for crack growth? It turns out that if the plastic zone is sufficiently small as compared to all of the length dimensions in the structure and test specimen, the stress intensity factor is still a valid fracture parameter that can uniquely characterize the condition for crack propagation, even if fracture actually nucleates inside the plastic zone.

This smallness requirement on the plastic zone restricts its existence well within the singularity-controlled region where Eqs. (2.47) to (2.52) are valid except for the plastic zone, as shown in Figure 2.9. Obviously, in a K-controlled stress field, it is always possible to associate the condition for crack propagation with a threshold value of K, whether the crack originates from the plastic zone or from the singularity-dominated region, implying the validity of Eqs. (2.77) to (2.79). If the plastic zone is not small compared with the characteristic length of the problem (such as the crack size), a singularity-dominated region where the stress varies as $1/\sqrt{r}$ no longer exists.

Consequently, the stress intensity factor loses its physical significance as a fracture parameter, and the stresses outside the plastic zone are described by Eq. (2.56) in which higher-order terms become significant. This implies that LEFM is no longer valid, and the problem has to be analyzed based on NLFM. The plastic zone theory will be discussed in the following section.

2.4 PLASTIC ZONE THEORIES AT CRACK TIP

For metallic materials, yielding takes place as the near-tip elastic stresses reach the yield strength of the material, leading to the formation of a plastic zone in front of the crack. For small-scale yielding, the size of the plastic zone can be estimated by two methods: the Irwin stress-correction approach and the Dugdale-Barenblatt cohesive zone models. While the former is based on simple corrections of the linear elastic solutions, the latter are nonlinear fracture mechanics models that focus on the fracture mechanism inside the plastic zone and can be extended to study fracture phenomena in other engineering materials that involve intense crack-tip inelasticity, such as concrete.

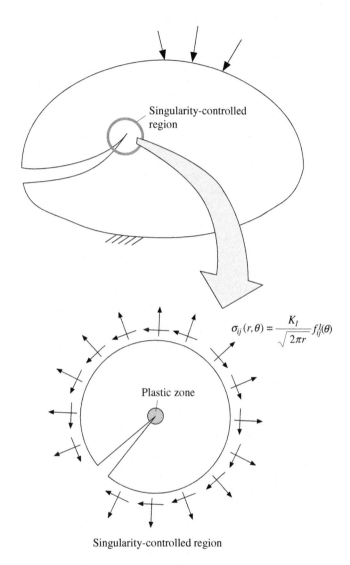

$$\sigma_{ij}(r,\theta) = \frac{K_I}{\sqrt{2\pi r}} f_{ij}^I(\theta)$$

FIGURE 2.9 Schematic illustration of K-controlled near-tip stresses with the presence of a plastic zone.

2.4.1 **The Irwin Plastic Zone Corrections**

Figure 2.10a illustrates the near-tip stress distribution σ_y (r, θ) in the plane $\theta = 0$, where $\sigma_y(r) = K_I/\sqrt{2\pi r}$ (refer to Eq. (2.47)). For plane stress, the size of the plastic zone can be simply estimated by enforcing the yield strength σ_{ys} on the principal stress $\sigma_y(r)$ such that $\sigma_y(r) = \sigma_{ys}$. Solving the equation for r leads to a first-order estimate:

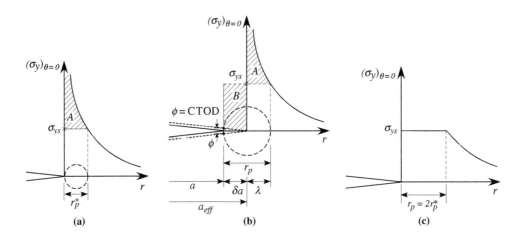

FIGURE 2.10 Irwin's plastic zone correction: (a) first-order estimate of size of plastic zone, (b) second-order estimate of size of plastic zone, and (c) elastic–plastic stress distribution at the crack tip.

$$r_p^* = \frac{K_I^2}{2\pi\sigma_{ys}^2} = \frac{\sigma^2 a}{2\sigma_{ys}^2} \tag{2.80}$$

It is noted that in the preceding derivation, redistribution of the stresses in the shaded area A above the yield strength is ignored. As a result, Eq. (2.80) is an underestimate of the plastic zone size. In reality, when yielding takes place, stresses must redistribute to satisfy the force equilibrium condition, resulting in a plastic zone larger than the size defined by Eq. (2.80). Irwin (1958, 1960) proposed a method to estimate this increase by defining an effective crack whose length is slightly longer than the actual crack, as illustrated in Figure 2.10b. As shown, the small increase of crack size is denoted by δa, and the effective crack is defined as $a_{eff} = a + \delta a$. Notice that the incremental crack δa carries the yield stress σ_{ys} over its entire length, and the total load carried is represented by the shaded area B. Based on the first-order estimate, the size of the plastic zone ahead of the effective crack is

$$\lambda = \frac{\sigma^2(a + \delta a)}{2\sigma_{ys}^2} \cong r_p^* \tag{2.81}$$

The remaining task is to find a value for δa so that the unaccounted load of A is completely redistributed into B—that is, $A = B$. This requirement can be expressed as

$$(\delta a + \lambda) \cdot \sigma_{ys} = \int_0^\lambda \sigma \sqrt{\frac{a + \delta a}{2r}}\, dr \tag{2.82}$$

Solving Eq. (2.82) for δa by using Eq. (2.81) and neglecting δa as compared to a, it turns out that

$$\delta a = r_p^* \text{ or } r_p = \delta a + \lambda = 2r_p^* = \frac{K_I^2}{\pi\sigma_{ys}^2} \tag{2.83}$$

Thus, the size of the plastic zone, r_p, based on the second-order estimation is twice as large as the first-order estimation, r_p^*. For an ideal elastic-plastic material the resulting stress distribution at the crack tip is illustrated in Figure 2.10c. Based on the Irwin plastic zone correction, r_p^*, the effective stress intensity factor is obtained as

$$(K_I)_{eff} = \sigma\sqrt{\pi a_{eff}} = \sigma\sqrt{\pi\left(a + r_p^*\right)} = \sigma\sqrt{\pi a\left(1 + \frac{\sigma^2}{2\sigma_{ys}^2}\right)} \tag{2.84}$$

In plane strain, the plastic zone is formed under the triaxial stress condition, and the Irwin plastic zone correction becomes

$$r_p^* = \frac{K_I^2}{6\pi\sigma_{ys}^2} \tag{2.85}$$

which is smaller than its counterpart in plane stress by a factor of 3.

Based on the effective crack concept, a useful expression for the crack-tip-opening displacement (CTOD), denoted as ϕ in Figure 2.10b, is derived. Notice that for plane stress, the crack-opening displacement (COD) in the central crack problem of Figure 2.4 is given by

$$COD = 2u_y = \frac{4\sigma}{E}\sqrt{a^2 - x^2} \tag{2.86}$$

For an ideally elastic situation, the CTOD is zero at $x = a$. Using the concept of the Irwin effective crack, however, ϕ is obtained as

$$\phi = \frac{4\sigma}{E}\sqrt{\left(a + r_p^*\right)^2 - a^2} = \frac{4K_I^2}{\pi E\sigma_{ys}} \tag{2.87}$$

The usefulness of this expression will be demonstrated in Chapter 3 when we examine the validity of a numerical solution obtained from a singularity-contained stress field in an elastic body.

2.4.2 Cohesive Zone Models by Dugdale and Barenblatt

Based on insightful analyses of the inelastic material behavior at the crack tip, Dugdale (1960) and Barenblatt (1959, 1962) proposed separately the concept of a cohesive zone model for studying plasticity at the crack tip. As shown in Figure 2.11, they also considered an effective crack of length $2a + 2\rho$; along the plastic zone ρ cohesive forces representing the material resistance to fracture are applied as external forces to close the crack so that the stress singularity disappears.

In Dugdale's model the closure stress is the yield strength σ_{ys}; in Barenblatt's model it represents the molecular force of cohesion and varies along the plastic zone. These models take nonlinear material behavior at the crack tip into account, and by introducing the cohesive forces directly to the crack surface, a fracture problem is thus defined with all the forces involved, including the transitory stress state at the crack tip where fracture originates. In the following, a

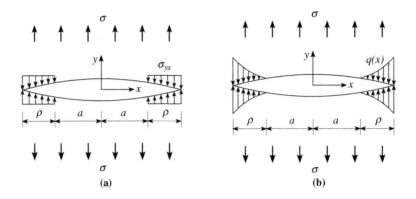

FIGURE 2.11 Dugdale and Barenblatt cohesive zone models: (a) Dugdale cohesive model, and (b) Barenblatt cohesive model.

relation between the applied load and the size of plastic zone is obtained by using the Dugdale model, which offers a closed-form solution due to the uniform distribution of closure stresses.

For the tip stresses of the effective crack to be bounded, the stress singularity must disappear under the combined actions of far-field tension and closure stresses at the crack edge, as shown in Figure 2.11a. Based on the principle of superposition, the stress intensity factor can be obtained by adding the stress intensities obtained separately from the two load components. For a central crack under remote tension, the stress intensity factor is given by

$$K_I^{(1)} = \sigma\sqrt{\pi(a+\rho)} \tag{2.88}$$

Figure 2.12 illustrates a pair of normal forces applied at an arbitrary location of a crack (crack length $= 2a$), and the resulting stress intensity factors at the two crack tips are given by

$$K_{I(+a)} = \frac{P}{\sqrt{\pi a}}\sqrt{\frac{a+x}{a-x}} \tag{2.89a}$$

$$K_{I(-a)} = \frac{P}{\sqrt{\pi a}}\sqrt{\frac{a-x}{a+x}} \tag{2.89b}$$

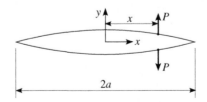

FIGURE 2.12 Wedge forces applied to open the crack.

Using these relations, the stress intensity factor due to the closure stresses is obtained by adding the contribution from both sides:

$$K_I^{(2)} = -\frac{\sigma_{ys}}{\sqrt{\pi(a+\rho)}} \int_a^{a+\rho} \left\{ \sqrt{\frac{a+\rho+x}{a+\rho-x}} + \sqrt{\frac{a+\rho-x}{a+\rho+x}} \right\} dx$$
$$= -2\sigma_{ys} \sqrt{\frac{a+\rho}{\pi}} \arccos\left(\frac{a}{a+\rho}\right) \tag{2.90}$$

Notice that the negative sign in Eq. (2.90) is due to the fact that the yield stress tends to close the crack. The superposition of the two stress intensity factors must satisfy the following condition:

$$K_I = K_I^{(1)} + K_I^{(2)} = 0 \tag{2.91}$$

Therefore,

$$\frac{a}{a+\rho} = \cos\left(\frac{\pi\sigma}{2\sigma_{ys}}\right) \text{ or } \rho = a\left[\sec\left(\frac{\pi\sigma}{2\sigma_{ys}}\right) - 1\right] \tag{2.92}$$

Equation (2.92) predicts large-scale yielding if the far-field stress approaches the yield strength. On the other hand, if the applied load is far below the yield stress, yielding is then confined to only a small area at the crack tip. Neglecting the higher-order terms in the series expansion of Eq. (2.92), the size of plastic zone is obtained as

$$\rho = \frac{\pi^2\sigma^2 a}{8\sigma_{ys}^2} = \frac{\pi K_I^2}{8\sigma_{ys}^2} \tag{2.93}$$

where $K_I = \sigma\sqrt{\pi a}$. It is interesting to compare Dugdale's estimation of the plastic zone size ρ with Irwin's second-order estimation r_p as defined in Eq. (2.83):

$$\frac{r_p}{\rho} = \frac{8}{\pi^2} = 0.81 \tag{2.94}$$

As seen, the relative difference between the two is less than 20 percent for small-scale yielding. Since the two estimates are obtained with quite different theoretical bases, this result is remarkable indeed. Because the Irwin approach is based purely on the force equilibrium in a near-tip elastic stress field, this reasonable agreement with the Dugdale approach, which is a nonlinear fracture mechanics model, confirms the earlier statement that for small-scale yielding, near-tip solutions of LEFM are still valid except for the plastic zone.

2.5 FRACTURE PROCESS ZONE AND TENSION-SOFTENING PHENOMENON IN CONCRETE

Concrete is a heterogeneous material that consists of aggregates and cement pastes bonded together at the interface, and the material is inherently weak in tension due to the limited bonding strength and various preexisting microcracks and flaws that form during hardening of the matrix. The tensile strength of concrete approximately ranges from 8 to 15 percent of its compressive strength. Under external loading, a tension zone forms near the crack tip, in which complicated microfailure mechanisms take place. These fracture processes include microcracking, crack deflection, crack branching, crack coalescence, and debonding of the aggregate from the matrix, which are examples of inelastic toughening mechanisms that coexist with a crack when it propagates. In concrete, the inelastic zone at the crack tip is extensively developed and therefore, in principle, LEFM cannot be used to study the fracture of concrete.

Figure 2.13 schematically illustrates the formation of an inelastic zone in concrete, which is known as a fracture process zone (FPZ) that can be roughly divided into a bridging zone and a microcracking zone, along with two idealizations of the FPZ. It is known that bridging is a result of the weak interface between the aggregates and the cement pastes, and it is an important toughening mechanism in concrete. Within the damage zone the effective modulus of elasticity is reduced from that of the undamaged material E to E^*, if the process zone is modeled as a region of strain softening as shown in Figure 2.13b.

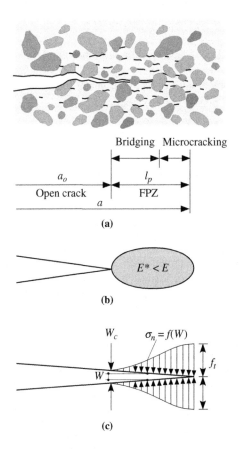

FIGURE 2.13 Concept of FPZ and tension-softening in concrete: (a) FPZ in front of an open crack, (b) reduced effective modulus of elasticity inside FPZ, and (c) tension-softening inside FPZ.

Inspired by the concept of the Dugdale and Barenblatt cohesive zone models, Hillerborg et al. (1976) envisioned a fictitious crack in place of the physical FPZ and subjected it to closure tractions, as shown in Figure 2.13c. As is illustrated in the figure, the closure stress associated with the bridging grains and microcracks is a maximum f_t at the tip of the FPZ and decreases to zero at the continuous crack tip where the COD reaches its critical value W_c, beyond which an open crack forms. Known as the tension-softening phenomenon, the relation between the closure stress and the COD with which the fracture energy of concrete is completely defined describes the local material behavior inside the FPZ when fracture takes place in concrete.

Figure 2.14 presents a numerically obtained load-deformation relation of a laboratory size, notched plain concrete beam under bending, and the growth of an FPZ at the notch tip based on the fictitious crack model by Hillerborg et al. The correspondence between the various points of the figure is meant to convey a clear picture on the FPZ and how it develops in the process of beam failure. As the figure shows, in the prepeak region the tip stress at the notch reaches the tensile strength of concrete at point A, signaling the initiation of an FPZ. Upon reaching the peak load at point B, the FPZ has grown to a length of 50 mm, and the tensile stress at the notch tip decreases to $0.55\,f_t$. The tip stress of the notch drops to zero at point C in the postpeak region, as the FPZ stretches to its full length of 150 mm, reaching approximately three-fourths of the beam height. Beyond point C, a completely open crack appears ahead of the notch and continues to extend until the beam breaks into two at point D. The dashed line drawn across the load-deformation curve divides the structural response into two regions.

As seen, the prepeak nonlinearity and the tension-softening in the pre- and the early postpeak regions (somewhere above point C) are mainly the work of microcracking. The remaining structural response before final failure at point D is the result of the fracture processes that involve mainly aggregate interlock and other frictional effects.

As clearly shown in Figure 2.14, with the presence of such a large FPZ at the notch tip, the fracture of concrete cannot be described by a single material parameter, such as K_{Ic} or G_{Ic}, as in the case of brittle materials. The fracture behavior and structural response of the beam can only be described when the FPZ and the tension-softening characteristics are taken into account based on the principles of NLFM of concrete, which will be discussed following.

2.6 FRACTURE ENERGY G_F AND TENSION-SOFTENING LAW IN CONCRETE

As just discussed, fracture of concrete initiates in the FPZ ahead of an open crack through complicated microfailure processes, and the fracture energy is consumed in overcoming the resistance of various toughening mechanisms to form an open crack at the end of the FPZ. The amount of fracture energy required to break a unit area of concrete is generally regarded as a material property (although it varies slightly with size) that determines the fracture behavior of the material through the fundamental relationship between the cohesive stress and the COD in the FPZ, which is known as the tension-softening law of concrete.

Just like the constitutive relationship of a continuous material that stipulates the fundamental material behavior (whether it is elastic or inelastic), the tension-softening law with the fracture energy as its defining characteristic is the constitutive relationship for the material in the FPZ that

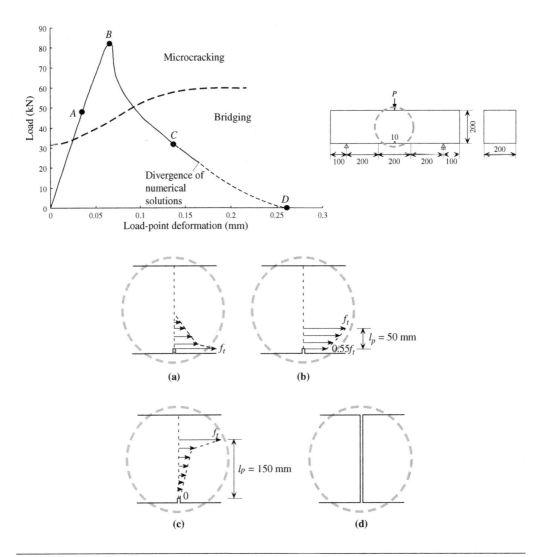

FIGURE 2.14 Typical load-deformation relation of a notched beam under bending and development of the FPZ in front of the notch (dimensions in mm).

describes the transitional material behavior from the continuous state to the discontinuous state—in other words, how the tensile stress decreases with the increasing discontinuity in the FPZ. In the following, the fracture energy is discussed through the numerical example of Figure 2.14 and the standard RILEM (1985) test method for its measurement. Tension-softening relations are then introduced through experimental and numerical studies.

2.6.1 **Fracture Energy G_F**

In the load-displacement relations shown in Figure 2.14, the area enclosed by the response curve and the horizontal axis represents the work done by the external load to fracture the beam. Suppose that the crack growth is stable and the work done by the external load is spent solely in crack propagation. Based on the Griffith energy criterion, crack growth in an elastic body in the equilibrium state is a natural process of energy transfer between the strain energy of the body and the fracture energy required for creating a new crack surface so that a state of minimum potential energy is achieved for the system at a given load level. In the present case, the work is consumed in breaking the unnotched part of the beam cross-section—the ligament ahead of the notch. Denoting the work of the external load by W_F and the ligament area by A_{lig}, the energy needed to create a crack of unit area, G_F, is obtained as

$$G_F = \frac{W_F}{A_{lig}} \tag{2.95}$$

In fracture mechanics of concrete, G_F represents the mode-I fracture energy, which is often simplified as fracture energy.

As mentioned before, at point C of the response curve in Figure 2.14, the FPZ has grown to its full length with the COD of the fictitious crack at the end of the FPZ reaching its critical value W_c, with which the tensile stress decreases to zero and an open crack appears. Along this FPZ a full relationship between the cohesive stress and the COD is obtained and plotted in Figure 2.15, which is known as the tension-softening relation of concrete. Obviously, this relationship can also be

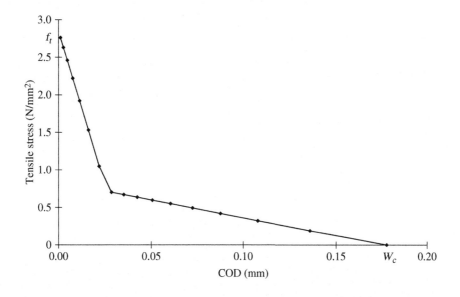

FIGURE 2.15 Relations between the tensile stress and the COD along the FPZ at point C of the response curve in Figure 2.14.

obtained from the stress-COD relations at the notch tip, where an open crack has just been created by fracturing the intact material there. Notice that the area enclosed by this tension-softening curve with the horizontal axis in Figure 2.15 is exactly the fracture energy defined by Eq. (2.95)—that is,

$$G_F = \int_0^{W_c} \sigma_n(W)\,dW \tag{2.96}$$

In 1985, the RILEM Technical Committee 50-FMC (Fracture Mechanics of Concrete) proposed a test method for determining the fracture energy G_F of mortar and concrete by means of stable three-point bend tests on notched beams (RILEM, 1985), as shown in Figure 2.16a. The size of the beam depends on the maximum size of aggregate, d_a, according to Table 2.1. As seen

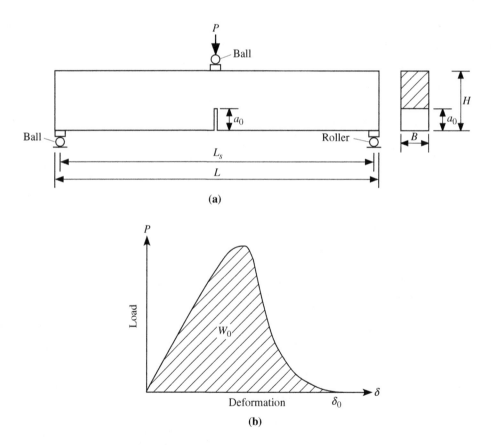

FIGURE 2.16 Determination of fracture energy G_F based on the RILEM method: (a) notched beam under three-point bending, and (b) load-deformation relations.

Table 2.1 Recommended Sizes of Beams for Measuring G_F

d_a (mm)	Depth, H (mm)	Width, B (mm)	Length, L (mm)	Span, L_s (mm)
1.0–16	100 ± 5	100 ± 5	840 ± 10	800 ± 5
16.1–32	200 ± 5	100 ± 5	1190 ± 10	1130 ± 5
32.1–48	300 ± 5	150 ± 5	1450 ± 10	1385 ± 5
48.1–64	400 ± 5	200 ± 5	1640 ± 10	1600 ± 5

d_a: Maximum size of aggregate
Source: RILEM (1985).

from the table, the depth of the beam is at least six times the aggregate size, and the span-to-depth ratio can vary between 4 and 8. To ensure that the work performed by the external load will be spent entirely in fracturing the beam at the notch, a notch size as large as half the beam depth is used. With such a large notch at the midspan, the maximum bending moment can be expected to be kept reasonably low to ensure that all deformations outside the fracture zone are purely elastic due to the low stresses there.

To ensure stable crack propagation during the test, the testing machine must be stiff enough or furnished with a closed-loop servo control. Recommendations with regard to the stiffness of the machine are given in RILEM (1985). Notice that a test is regarded as stable if the load and the deformation change slowly during the whole test, without any sudden jumps. To ensure pure bending, the supports and loading arrangements should be arranged in such a manner that the forces acting on the beam are statically determined, as shown in Figure 2.16a. During the test, the vertical displacement at midspan is measured and continuously plotted along with the gradually applied load P until complete fracture of the specimen. As shown in Figure 2.16b, the area under the load-deformation curve, W_0, represents the work done by the external load to fracture the beam. Besides this major work of fracture, there are other types of loads that also perform work of fracture that is secondary but may not be negligible. These include the self-weight of the specimen and the loading equipment carried by it. Therefore, the fracture energy should be calculated as

$$G_F = \frac{W_F}{A_{lig}} = \frac{W_0 + mg\delta_0}{(H - a_0)B} \tag{2.97}$$

where W_F is the total work done by all the loads involved, mg is the weighted sum of the loads that perform the secondary work, and δ_0 is the displacement at failure.

From a fracture mechanics point of view, the RILEM method is conceptually clear, simple, and accurate as a test method for measuring the fracture energy of mortar and concrete. From an experimental point of view, the fracture test on a three-point bend beam is both practical and reliable. For these reasons the principles embodied in the RILEM method still serve as the guidelines for designing test methods for the measurement of this fracture-related material constant for various structural concrete.

2.6.2 **Tension-Softening Law**

As shown in Figure 2.15, the tension-softening relation of concrete possesses two distinctive features: the steep descending slope caused by the rapid loss of tensile strength in the initial stage of softening and a long tail with the increasing crack-opening displacement, illustrating the persistent stress-transferring capability of aggregate interlocking in the FPZ. Due to the displacement discontinuity in the FPZ and local inhomogeneities of the concrete material, establishing the exact relations between the decreasing tensile stress and the increasing COD through experiments (e.g., the uniaxial tension test) is a demanding task that requires both rigorous testing methods and sophisticated skill and experience in carrying out the tests and processing test data to derive these relations. Through the great efforts of many researchers, the concept of the tension-softening law of concrete has been firmly established as the basis for fracture mechanics of concrete, and the general shape of this constitutive relation has been obtained using deformation-controlled uniaxial tension tests as well as other creative methods, such as the inverse modeling method.

Uniaxial Tension Tests

Because the experimental investigation carried out by a group of scientists in the Stevin Laboratory of the Delft University of Technology in the 1980s is widely recognized as one of the most rigorous approaches on this subject, their test results are briefly introduced here. For details of these experimental studies, refer to Cornelissen et al. (1986) and Hordijk (1991). The schematic test setup for the deformation-controlled uniaxial tension test is shown in Figure 2.17, along with one type of measuring device and the arrangement on the test specimen. As seen, a narrow prismatic test specimen of size $60 \times 50 \times 150$ mm^3 was glued to the loading platens, and two saw cuts were introduced into the specimen to reduce the middle cross-sectional area to 50×50 mm^2. To minimize the nonuniform crack openings at the notched cross section, a guiding system was designed and connected to the top loading platen to prevent it from rotating to make sure that the top and the bottom loading platens remained parallel throughout the test.

To prevent any frictional force of this guiding system from entering the load reading, the load cell was moved from above the guiding system to beneath the bottom loading platen as shown in Figure 2.17a, knowing that even a small frictional force (estimated to be only 3 percent of the tensile strength of concrete in these tests) could significantly influence the tail shape of the softening curve. The closed-loop servo-hydraulic loading machine was controlled at a deformation rate of 0.08 μm/s corresponding to a measuring length of 50 mm. Deformation of double-edge notched specimens was measured using the average signal of the deformations at the four corners of the specimen recorded by four linear variable differential transducers (LVDT). For the deformation control, the signals for the LVDTs were averaged, and the average signal was compared with the signal from a specially designed ramp generator to eliminate noise.

It is known that during a tension test the FPZ does not establish itself instantaneously right across the specimen as the initial microcracking activity is dispersed; the deformation of the specimen only becomes localized in an FPZ after the maximum load has been reached. Since the crack opening cannot be measured directly, it must be derived from the total deformation δ that is found for the applied measuring length by deducting the elastic deformation δ_e and any prepeak inelastic deformation δ_{ine}, according to the procedure shown in Figure 2.18.

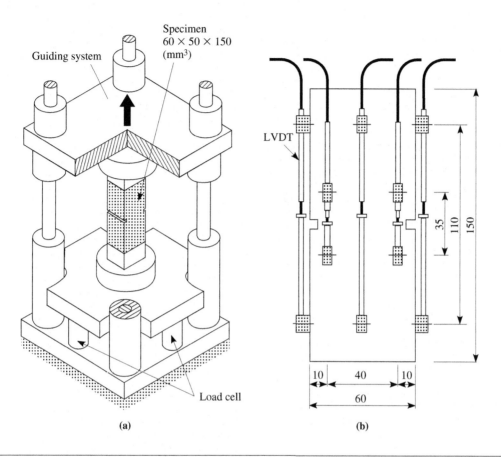

FIGURE 2.17 Schematic representation of a tensile specimen glued in the testing equipment and one type of measuring device applied (dimensions in mm): (a) uniaxial tension and (b) measuring devices (after Hordijk, 1991).

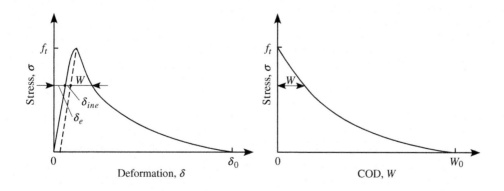

FIGURE 2.18 Procedure to derive a σ-w relation from a σ-δ relation, obtained from a deformation-controlled uniaxial tensile test.

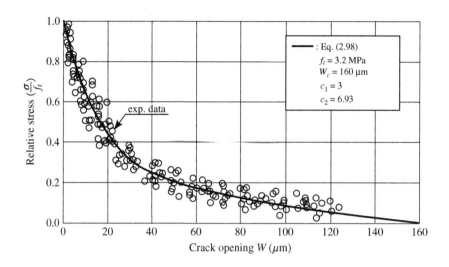

FIGURE 2.19 Regression analysis for the σ-w relation; data points and fitted relation (after Hordijk, 1991).

Figure 2.19 presents the experimental results obtained from a series of tests (details of the testing equipment and specimen dimensions were slightly different from those shown in Figure 2.17) on normal-weight concrete ($\gamma = 2370$ kg/m³, $f_c = 47$ MPa, $E = 39$ GPa, and $f_t = 3.2$ MPa), where the data points of individual experiments represent the relation between the normalized stress, σ/f_t, and the COD, W. A regression analysis on these test data was carried out, and the optimum relation (standard deviation = 0.052) was given by

$$\frac{\sigma}{f_t} = \left[1 + \left(c_1 \frac{W}{W_c}\right)^3\right]\exp\left(-c_2 \frac{W}{W_c}\right) - \frac{W}{W_c}\left(1 + c_1^3\right)\exp(-c_2) \tag{2.98}$$

The best-fit curve was obtained for $c_1 = 3$, $c_2 = 6.93$, and $W_c = 160$ μm, and the corresponding fracture energy was found to be $G_F = 0.1$ N/mm based on Eq. (2.96), resulting in a relation of $W_c = 5.14\ G_F/f_t$.

Hordijk (1991) made the following two comments concerning the tail portion of the curve.

1. Only data points up to an arbitrarily chosen crack opening of 120 μm, where the transferable stress was already diminished to about 10 percent of f_t, were used in the numerical regression analysis. If more data points for larger crack openings were taken into account, this would give too much weight to that part of the curve in the regression analysis, which would result in a less acceptable fit for the smaller crack openings.

2. The function (Eq. 2.98) was chosen in such a way that there is an intersection point W_c (critical crack opening) with the horizontal axis ($\sigma/f_t = 0$), which corresponds to the real material behavior where the crack surfaces are completely separated at a certain crack opening. Although there is a correlation between W_c obtained from the regression analysis and the real maximum crack opening W_o, this does not mean that the value as found previously ($W_c = 160$ μm) coincides with the real value for W_o. In Eq. (2.98), W_c is more or less used as a fitting parameter.

Considering the fact that the bulk portion of fracture energy is consumed in the range of the data employed for the regression analysis, these comments are deemed relevant because the fracture energy spent in the region of small CODs has a much stronger influence on fracturing behavior than it does in the region of large CODs, as will be clearly shown through numerical studies in the later chapters of the book.

Next, a comparison study was carried out based on the results of a number of experimental studies using uniaxial tension tests (with one exception) as reported in the literature, focusing on whether the obtained tension-softening relations in Figure 2.19 are of a representative nature. The sources of the literature are listed in Table 2.2. In Figure 2.20a the stress is normalized by the tensile strength f_t, and the experimental results are compared with the obtained scatter band in Figure 2.19. As shown, most of the curves fit in this scatter band. It was found that the few curves that fell out of the envelope belonged to particular types of concrete mixes, such as those with rather low compressive strength (curve 3), limestone aggregates (curve 6), or lightweight concrete (curve 8). Notice that in these types of concrete (lightweight or limestone concrete) the fracture mechanism is different from the ordinary type of concrete—for example, cracks running through the aggregates rather than the aggregates being pulled out of the cement paste. The close correlations of the curves in Figure 2.20a for those with normal concrete suggest that with f_t and the shape function of Eq. (2.98), a rather good approximation of the tension-softening curve for an ordinary type of concrete can be obtained.

To improve the description of the tension-softening relation, the fracture energy should be employed explicitly as another parameter. In Figure 2.20b the nondimensionalized tension-softening curves are presented, with the COD being normalized by its critical value W_c, which is defined as $W_c = 5.14G_F/f_t$ based on the best-fitting curve of Eq. (2.98). The fracture energy for the tests listed in Table 2.2 was calculated according to the approximations as defined in the inset of Figure 2.20b. As can be seen, after normalizing the COD, all the curves fit into a narrow band, including those representing deviating concrete mixes. Since these tests were carried out by different research groups independently, using different concrete mixes, specimen dimensions, testing equipment, and even different testing methods, the close agreement of these test results clearly shows that if f_t and G_F are known, the tension-softening relation can be reasonably approximated by Eq. (2.98) in which $c_1 = 3$, $c_2 = 6.93$, and $W_c = 5.14G_F/f_t$.

Inverse Modeling Method

Because of the difficulty in carrying out a stable uniaxial tension test as just described, Roelfstra and Wittmann (1986) proposed an inverse modeling method for determining the tension-softening relation using the fictitious crack approach and data fitting. In this method a conventional fracture test (such as a bending test) is simulated numerically by assuming and modifying a bilinear tension-softening relation that contains four parameters, as shown in Figure 2.21, through iterative computations to reduce the differences between the experimental and numerical results. They found that the best-fitting response curve (such as a load-deflection curve) could be obtained when a good approximation to the actual tension-softening characteristics was reached. The original formulation has been modified by other researchers to improve the convergence of solutions, and the revised formulation by Nomura et al. (1990) is outlined following.

Table 2.2 Information About Deformation-Controlled Uniaxial Tensile Tests for Which $\sigma\text{-}W$ Relations Are Compared

No	Reference	Dimensions[a] (mm³)	l_{meas} (mm)	f_c (MPa)	$f_{t,spl}$ (MPa)	w/c	d_a (mm)	age[b] (days)
1	Petersson (1981)	50 × 30 × 20	40	—	—	0.50	8	28
2	Notter (1982)	500 × 125 × 50	100	44	—	0.50	16	28
3	Gylltoft (1983)	85 × 30 × 30	67	28	2.8	—	—	—
4	Reinhardt and Cornelissen (1984)	300 × φ 120	25	45	2.9	0.50	16	32
5	Eligehausen and Sawade (1985)	600 × φ 100	100	35	3.0	0.76	16	150
6	Gopalaratnam and Shah (1985)[c]	305 × 60 × 19	85	44	—	0.45	10	28
7	Körmeling (1986)	100 × φ 74	100	44	3.5	0.50	8	—
8	Hordijk et al. (1987)[d]	150 × 50 × 50	35	50	3.4	0.59	8	150
9	Guo and Zhang (1987)	210 × 70 × 70	155	31	2.4	0.60	—	33
10	Scheidler (1987)	450 × 300 × 150	450	—	—	0.70	16	28
11	Wolinski et al. (1987)	150 × 50 × 50	35	47	3.1	0.50	8	32
12	Rokugo et al. (1989)[e]	—	—	58	—	0.52	15	—

[a]Length × width × depth or length × diameter for the critical cross-sectional area
[b]Mean age in case of a range
[c]Crushed limestone for the bigger aggregates
[d]Lightweight concrete with sintered expanded clay 4–8 mm
[e]Tube tension test
Source: After Hordijk (1991).

58

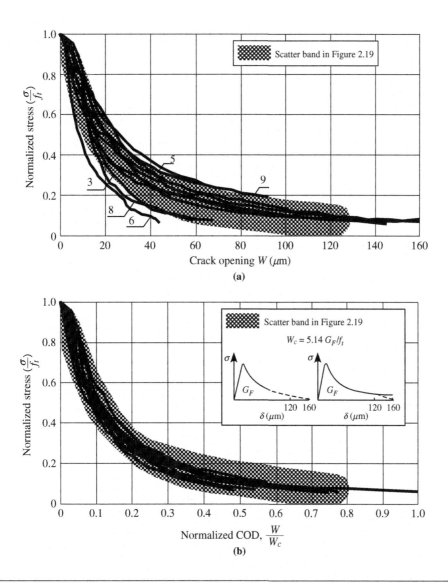

FIGURE 2.20 Comparisons of tension-softening relations obtained by different research groups: (a) $\sigma/f_t - W$ relations, and (b) $\sigma/f_t - W/W_c$ relations (after Hordijk, 1991).

Let $F_E(i)$ represent the experimentally obtained load-deformation curve recorded in a time sequence ($i = 1, 2, \ldots, r$) at a given location. The numerically calculated load-deformation curve at the same spot is denoted as $F_M(i; \{x\})$, which is obtained by assuming a bilinear softening relation with the four parameters $\{x\} = \{f_t, s_1, W_1, W_2\}$. By minimizing the following sum of the square differences of the two curves, the parameters for an optimum softening relation can be determined:

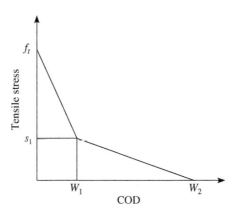

FIGURE 2.21 Definition of f_t, s_1, W_1, and W_2 in the bilinear softening diagram.

$$S(\{x\}) = \sum_{i=1}^{r} [F_E(i) - F_M(i; \{x\})]^2 \rightarrow \min \tag{2.99}$$

Let $\{x^{(k)}\}$ and $\{\Delta x^{(k)}\}$ represent, respectively, the kth estimation of the softening parameters and the corresponding incremental modification. The next estimation is then given by $\{x^{(k+1)}\}$ $= \{x^{(k)}\} + \{\Delta x^{(k)}\}$. The incremental modification $\{\Delta x^{(k)}\}$ is estimated from the following iterative equation:

$$\{\Delta y^{(k)}\} \cong [J^{(k)}]\{\Delta x^{(k)}\} \tag{2.100a}$$

where

$$\Delta y_i^{(k)} = F_E(i) - F_M\left(i; \{x^{(k)}\}\right) \tag{2.100b}$$

$$J_{ij}^{(k)} = \left[\frac{\partial F_M(i; \{x\})}{\partial x_j}\right]_{x_j = x_j^{(k)}} \tag{2.100c}$$

The flow chart for calculating the parameters of an optimum softening relation is illustrated in Figure 2.22. For details of the solution procedure, refer to Nomura et al. (1990).

Hordijk compared the experimental results presented in Figure 2.20 with the optimum bilinear relations obtained by Wittmann et al. (1988) for a number of concrete mixes and loading rates, in which the stress at the break-point was fixed at ft/4, and he found good agreement between them. He observed that these optimum relations with only limited differences from each other appear to be a good approximation of the directly determined softening curve given by Eq. (2.98), as shown in Figure 2.23, where the averaged bilinear relation is presented. As seen, both the initial slope and the slope of the second part of the descending branch coincide very well, while the parameter W_c is found to be coincidentally the same.

The success of the inverse modeling method just discussed highlights the relevance of adopting a bilinear relation to represent the real tension-softening characteristics of concrete, which,

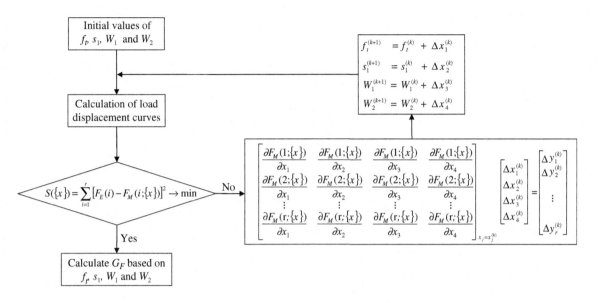

FIGURE 2.22 Flowchart of the numerical procedure for inverse modeling.

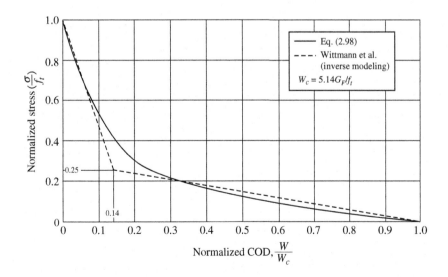

FIGURE 2.23 Bilinear softening relation by inverse modeling as compared with the relation obtained from uniaxial tensile tests (after Hordijk, 1991).

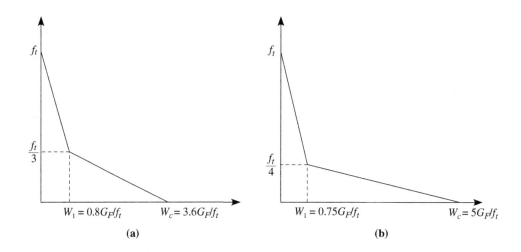

FIGURE 2.24 Proposed bilinear tension-softening models (a) by Petersson (1981) and (b) by Rokugo et al. (1989).

according to Cotterell and Mai (1996), "has the advantage of simplicity and has enough parameters to enable the load-deflection curve to be accurately predicted by numerical simulation." Figure 2.24 presents two types of bilinear relation that have been proposed for a wide range of concrete. Petersson (1981) found that the parameters of $W_1 = 0.8G_F/f_t$, $W_c = 3.6G_F/f_t$, and $s_1 = f_t/3$ could lead to a good fit to the tension-softening curve for a range of concrete.

Based on the optimum bilinear relations obtained by Wittmann et al. (1988; Figure 2.23), Rokugo et al. (1989) modified the parameters slightly and suggested a bilinear relation with $W_1 = 0.75G_F/f_t$, $W_c = 5G_F/f_t$, and $s_1 = f_t/4$. Realistic predictions of fracture processes and structural responses in various concrete structures have been obtained by using these models. When these proposed relations fail to represent the real tension-softening characteristics for a particular type of concrete, the inverse modeling method can be employed to obtain an optimum softening relation for that particular material. With a bilinear softening relation, the mode-I fracture energy is given by

$$G_F = \frac{1}{2}(f_t W_1 + s_1 W_c) \tag{2.101}$$

REFERENCES

Anderson, T. L. (2005). *Fracture Mechanics—Fundamentals and Applications*, 3rd ed., Taylor and Francis.

Barenblatt, G. I. (1959). "The formation of equilibrium cracks during brittle fracture: general ideas and hypotheses, axially-symmetric cracks." *J. Appl. Math. Meth.* 23, 622–636.

Barenblatt, G. I. (1962). "The mathematical theory of equilibrium cracks in brittle fracture." *Advances in Appl. Mech.*, 7, 55–129.

Broek, D. (1986). *Elementary Engineering Fracture Mechanics*, 4th ed., Kluwer Academic.

Cornelissen, H. A. W., Hordijk, D. A., and Reinhardt, H. W. (1986). "Experimental determination of crack softening characteristics of normal-weight and light-weight concrete." *HERON*, 31(2), 45–56.

Cotterell, B., and Mai, Y. W. (1996). *Fracture Mechanics of Cementitious Materials*, Blackie Academic & Professional.

Dugdale, D. S. (1960). "Yielding of steel sheets containing slits." *J. Mech. Phys. Solids*, 8, 100–104.

Eligehausen, R., and Sawade, G. (1985). "Verhalten von Beton auf Zug." *Betonwerk und Fertigteil-Technik*, 5, 315–322.

Fan, T. (2003). *Fundamentals of Fracture Theories*, Science Publishers, Beijing.

Gopalaratnam, V. S., and Shah, S. P. (1985). "Softening response of plain concrete in direct tension." *ACI Journal*, 82(3), 310–323.

Griffith, A. A. (1921). "The phenomena of rupture and flow in solids." *Phil. Trans. Royal Society*, Series A, 221, 163–198.

Griffith, A. A. (1924). "The theory of rupture." *Proc. 1st Intern. Congr. Appl. Mech., Delft*, 55–63.

Guo, Z., and Zhang, X. (1987). "Investigation of complete stress-deformation curves for concrete in tension." *ACI Journal*, 82(3), 310–324.

Gylltoft, K. (1983). *Fracture mechanics models for fatigue in concrete structures*, PhD Thesis, Lulea University of Technology.

Hillerborg, A., Modeer, M., and Petersson, P. E. (1976). "Analysis of crack formation and crack growth in concrete by means of fracture mechanics and finite elements." *Cement and Concrete Research*, 6 (6), 773–782.

Hordijk, D. A. (1991). *Local approach to fatigue of concrete*, PhD Thesis, Delft University of Technology.

Hordijk, D. A., Reinhardt, H. W., and Cornelissen, H. A. W. (1987). "Fracture mechanics parameters of concrete from uniaxial tensile tests as influenced by specimen length." In *Fracture of Concrete and Rock*, S. P. Shah and S. E. Swartz eds. Preprint SEM-RILEM Int. Conf., Bethel, 138–149.

Inglis, C. E. (1913). "Stresses in a plate due to the presence of cracks and sharp corners." *Trans. Inst. Naval Architects*, 55, 219–230.

Irwin, G. R. (1957). "Analysis of stresses and strains near the end of a crack traversing a plate." *J. Applied Mechanics*, 24, 361–364.

Irwin, G. R. (1958). "Fracture." *Handbuch der Physik VI*, S. Flugge ed., pp. 551–590, Springer.

Irwin, G. R. (1960). "Plastic zone near a crack and fracture toughness." *Mechanical and Metallurgical Behavior of Sheet Materials*, Proc. 7th Sagamore Conf. IV-63-IV-78.

Kormeling, H. A. (1986). *Strain rate and temperature behavior of steel fiber concrete in tension*, PhD Thesis, Delft University of Technology.

Muskhelishvili, N. I. (1953). *Some Basic Problems in the Theory of Elasticity*, Noordhoff.

Nomura, N., Mihashi, H., Suzuki, A. and Izumi, M. (1990). "Mechanism of brittleness in high-strength concrete based on nonlinear fracture mechanics." *J. Struct. Constr. Eng.* 416, 9–14.

Notter, R. (1982). *Schallemissionsanalyse fur Beton im dehnungsgesteuerten Zugversuch*, PhD Thesis, Zurich.

Petersson, P. E. (1981). "Crack growth and development of fracture zones in plain concrete and similar materials." *Report TVBM-1006*, Division of Building Materials, Lund Institute of Technology, Sweden.

Reinhardt, H. W., and Cornelissen, H. A. W. (1984). "Post-peak cyclic behavior of concrete in uniaxial and alternating tensile and compressive loading." *Cement and Concrete Res.*, 14, 263–270.

RILEM Draft-Recommendation (50-FMC). (1985). "Determination of the fracture energy of mortar and concrete by means of three-point bend tests on notched beams." *Materials and Structures*, 18, 285–290.

Roelfstra, P. E., and Wittmann, F. H. (1986). "Numerical method to link strain softening with failure of concrete." *Fracture Toughness and Fracture Energy of Concrete*, F. H. Wittmann ed., pp. 163–175, Elsevier Science B.V.

Rokugo, K., Iwasa, M., Suzuki, T. and Koyanagi, W. (1989). "Testing methods to determine tensile strain softening curve and fracture energy of concrete." *Fracture Toughness and Fracture Energy-test Methods for Concrete and Rock*, H. Mihashi, H. Takahashi, and F. H. Wittmann eds., pp. 153–163, Balkema Publishers.

Scheidler, D. (1987). "Experimentelle und analytische Untersuchungen zur wirklichkeitsnahen Bestimmung der Bruchschnittgrossen unbewehrter Betonbauteile unter Zugbeanspruchung." *ADfStb*, Heft 379, 94.

Williams, M. L. (1952). "Stress singularities resulting from various boundary conditions in angular corners of plates in extension." *J. Applied Mechanics*, 19, 526–528.

Williams, M. L. (1957). "On the stress distribution at the base of a stationary crack." *J. Applied Mechanics*, 24, 109–114.

Wittmann, F. H., Rokugo, K., Bruhwiler, E., Mihashi, H., and Simonin, P. (1988). "Fracture energy and strain softening of concrete as determined by means of compact tension specimens." *Materials and Structures*, 21, 21–32.

Wolinski, S., Hordijk, D. A., Reinhardt, H. W., and Cornelissen, H. A. W. (1987). "Influence of aggregate size on fracture mechanics parameters of concrete." *Int. J. Cement Composites and Lightweight Concrete*, 9(2), 95–103.

The Fictitious Crack Model and Its Numerical Implementation

3

3.1 INTRODUCTION

The presence of the fracture process zone (FPZ) in front of an open crack poses an analytical challenge in the framework of continuum solid mechanics for the study of concrete fracture because macroscopically this zone can be characterized neither as a continuous region nor as a discontinuous region. In fact, it is a partially damaged zone with some remaining stress-transferring capability through aggregate interlocking and various microcracking activities, and it functions like a transition zone between the open crack of complete discontinuity and the intact material of complete continuity beyond it. Since fracture of concrete originates in this zone, the analysis of cracks in concrete starts with the issue of how to model the FPZ.

Two concepts for modeling the FPZ have gained popularity in the development of computational theories for the fracture mechanics of concrete: the discrete-crack approach and the smeared-crack approach. In the discrete-crack approach, the FPZ is modeled as a fictitious crack that is subjected to external forces that are equivalent to the cohesive forces transferred through the FPZ to the surrounding elastic body. Because the forces exerted on the elastic zone by the fictitious crack remain unchanged, as with the presence of the physical FPZ, this approach is an accurate mathematical description of the problem. In the smeared-crack approach, a different modeling concept is employed, in which the localized inelastic deformations in the FPZ are smeared over a band of a certain width in terms of stress-strain relations, and thus the FPZ is modeled in a continuous fashion. In its finite element (FE) implementation, the material stiffness and strength of a cracked element are reduced according to a strain-softening relation, which is tantamount to smearing a discrete crack over an FE mesh or meshes to approximate the effect of material damage due to a single crack. Clearly, this method is an approximate approach to crack analysis.

As stated in Chapter 1, this book focuses on the fictitious crack model (FCM) and its theoretical extension to multiple-crack and mixed-mode fracture problems. In this chapter, the fundamental concept of the FCM and its numerical solution are discussed. Since its theoretical formulation is based on some important principles in linear elastic theory, these principles are reviewed. Also, the issue of stress singularity underlying the FCM formulation is addressed. Finally, some effective modeling techniques are introduced, including the dual-nodes method for crack-path modeling, the path-shifting method for remeshing of curvilinear crack trajectories, and incremental stress analysis.

65

3.2 HILLERBORG AND COLLEAGUES' FICTITIOUS CRACK MODEL

In this section the modeling concept of the FCM and its numerical formulation using the influence function will be explained.

3.2.1 Modeling Concept

As we saw in Figure 2.13, the FCM proposed by Hillerborg, Modeer, and Petersson (1976) for studying crack formation and crack growth in concrete brought two new concepts to the study of concrete fracture: the physical presence of an FPZ ("microcracked zone" in their original words) in front of an open crack due to strain localization and a constitutive law for crack propagation stipulating the relations between the cohesive forces and the crack-opening displacement (COD) inside the FPZ—in other words, the tension-softening relation. These are the fundamental concepts of the FCM that have become the basis of fracture mechanics of concrete.

Hillerborg and his coworkers developed these concepts based on experimental evidence of uniaxial tension tests on concrete bars that were performed by various researchers in the late 1960s (e.g., Hughes and Chapman, 1966; Evans and Marathe, 1968). These experimental studies showed that crack formation in concrete is an evolutionary process of strain buildup in the prepeak region, and strain localization and formation of a visible crack in the postpeak region, as illustrated in Figure 3.1. Therefore, in their modeling of concrete fracture they imagined the zone of strain localization as a fictitious crack that "corresponds to a microcracked zone with some remaining ligaments for stress transfer." They further suggested that this fictitious crack "may be looked upon as a reality. Stresses may be present in a microcracked zone as long as the

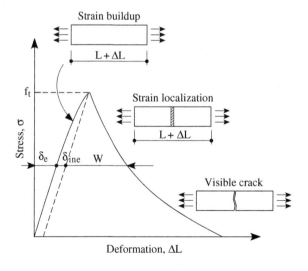

FIGURE 3.1 The evolution of crack formation.

corresponding displacement is small." As shown in Figure 3.1, for a bar of length L, the total deformation in the postpeak region is given by

$$\Delta L = \delta_e + \delta_{ine} + W \tag{3.1}$$

where δ_e is the elastic deformation, δ_{ine} is the inelastic deformation, and W is the COD of the fictitious crack. For the crack to propagate, the stress at the crack tip was assumed to reach the tensile strength of the material. They further assumed that the stress transferred through the fictitious crack was a function of the COD:

$$\sigma = f(W) \tag{3.2}$$

where $f(W)$ is a tension-softening curve unique to the material that must be determined from experiments, as discussed in the preceding chapter. As W reaches a critical value, W_c, the transferred stress drops to zero and the fictitious crack becomes a stress-free open crack.

For simplicity, Hillerborg and his coworkers further assumed that the inelastic deformation in the test specimen was negligible—that is, the behavior of the bulk material was linear elastic. Based on this assumption, Eq. (3.1) can be rewritten as

$$\Delta L = \delta_e + W = L\frac{\sigma}{E} + W = L\frac{f(W)}{E} + W \tag{3.3}$$

where E is the elastic modulus. Hence, if E and $f(W)$ are given, the load-deformation curve of the tensile bar is completely determined. As simple as Eq. (3.3) is, it underlines the general principles for crack analysis in concrete. First, it illustrates the fundamental importance of the tension-softening law: Without it, crack analysis in its strictest sense cannot be carried out. Second, it shows that after cracking, the general deformation of the cracked body is composed of two parts: the deformation of the elastic body and the crack-opening width of the crack. Third, the elastic body outside the cracking zone is subjected to the stress transferred through the fictitious crack, and therefore its deformation is also a function of the COD. As such, the essence of crack analysis is to determine the general deformation of the body after the occurrence of cracks. If the total deformation of the cracked body is determined, strain-and-stress analyses can then be carried out without any particular difficulty.

It should be noted that although the assumption of linear elasticity for the bulk material makes the problem easier to solve, it is not an essential requirement for crack analysis in concrete. As pointed out by several researchers, the FCM can also be applied to more general situations including nonlinear behavior of the bulk material and other cases (Elices and Planas, 1989; Planas et al., 1995). In general, however, the following hypotheses are used when applying the FCM in crack analysis of concrete:

1. The bulk material behavior is isotropic linear elastic, and thus it is defined by elastic modulus E, and Poisson's ratio, ν.

2. A crack initiates at a point when the maximum principal stress at that point reaches the tensile strength f_t, and the crack forms normal to the direction of the principal tensile stress.

3. As a crack forms and its crack opening increases, stress transfer takes place across the crack surfaces until a critical COD is reached. For a pure mode-I opening, the stress transferred between the surfaces of the crack is a function of the COD, as defined by Eq. (3.2).

It is obvious that Eqs. (3.1) and (3.3) are derived based on the known cracking behavior of a tensile bar, which is idealized as the crack forms perpendicular to the bar axis and its crack width is uniform. In most crack analysis problems, however, this is not the case, because the growth of a fictitious crack is a highly nonlinear process that usually requires numerical procedures to find the solution. A numerical formulation of a fictitious crack based on Petersson's influence function method will be introduced in the next section, which provides a computational means for determining the cohesive stress and shape of a fictitious crack.

3.2.2 **Numerical Formulation by Petersson's Influence Function Method**

In Petersson's formulation (1981) of the fictitious crack problem, a system of equations is established on the cracked cross section based on the principle of superposition in elasticity. By solving these equations, the unknown boundary conditions on the fictitious crack, including the cohesive stress and geometric shape of the crack, can be determined. Since the approach is similar to boundary integral methods, with the main difference being that the kernel of the integral equation is discretized a priori and determined by the finite element method (FEM), Bazant and Planas (1998) called it the pseudo-boundary-integral method.

State of the Problem

Consider a notched specimen that is symmetric to the crack plane and subjected to symmetric loading, as shown in Figure 3.2. Under the given load, a cohesive zone extends ahead of the notch

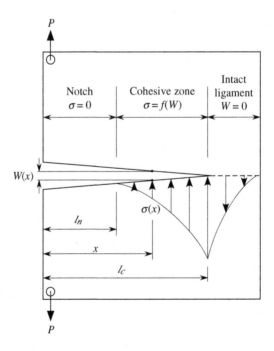

FIGURE 3.2 Schematic illustration of fictitious crack growth in pure opening mode.

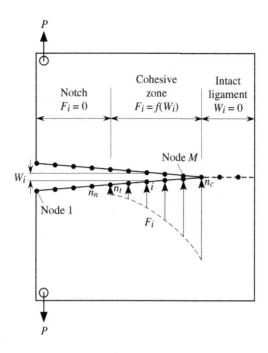

FIGURE 3.3 Node layout for the influence function method; the cohesive forces shown are those exerted on the lower half of the specimen.

tip as indicated in the figure. Along the surface of the fictitious crack, the following conditions must be satisfied:

$$\sigma(x) = 0 \quad \text{for } x < l_n \tag{3.4}$$

$$\sigma(x) = f[W(x)] \quad \text{for } l_n \leq x \leq l_c \tag{3.5}$$

$$W(x) = 0 \quad \text{for } x \geq l_c \tag{3.6}$$

where l_n is the notch size, l_c is the length of the fictitious crack, and x is the location of an arbitrary point along the crack.

Petersson's formulation is based on a discretization of the problem, as illustrated in Figure 3.3. As seen, the nodes that span the notch vary from 1 to n_n, with $n_t = n_n + 1$ denoting the tip of the notch (initial crack tip). The cohesive zone starts at $n_t \ (= n_n + 1)$ and ends at n_c; the nodes beyond n_c represent the uncracked ligament. Let M represent the total number of nodes on the fictitious crack; the node $M + 1$ then denotes the tip of the crack. Equations (3.4)–(3.6) are now written in terms of the nodal values as

$$F_i = 0 \quad \text{for } i = 1, \ldots, n_n \tag{3.7}$$

$$F_i = f(W_i) \quad \text{for } i = n_t, \ldots, n_c \tag{3.8}$$

$$W_i = 0 \quad \text{for } i \geq n_c \tag{3.9}$$

where F_i and W_i represent the nodal force and the COD at node i, respectively. Notice that a nodal force is the product of the nodal stress and the surface area apportioned to that node.

Numerical Formulation

Figure 3.4 illustrates the solution strategy proposed by Petersson in 1981 for analyzing the growth of a fictitious crack. As seen, the original problem in Figure 3.4a is treated as a composite loading case and is discomposed into a set of simple linear elastic problems, each with a single load condition. Based on the principle of superposition for linear elastic materials, the crack-opening width at node i of the fictitious crack is obtained by summing up the corresponding CODs that are linear functions of the external load and the closing forces at each node, as shown in Figures 3.4b and c, respectively. This can be written as

$$W_i = BK_i \cdot P + \sum_{j=1}^{M} AK_{ij} \cdot F_j \quad \text{for } i = 1, \ldots, M \tag{3.10}$$

where BK_i is the crack opening at node i produced by a unit external load and AK_{ij} is the crack opening at node i produced by a pair of closing forces applied at node j. The coefficients of BK_i and AK_{ij} can be obtained by linear elastic FE computations. By virtue of the reciprocity theorem, AK_{ij} is symmetric—that is, $AK_{ij} = AK_{ji}$. In Eq. (3.10), F_j and W_i are unknowns, and their solutions are subject to the conditions as specified in Eqs. (3.7) and (3.8). Obviously, with the fictitious crack now being described by $2M$ equations with $2M$ unknowns, a solution can be sought for a given load P, since these equations are linearly independent.

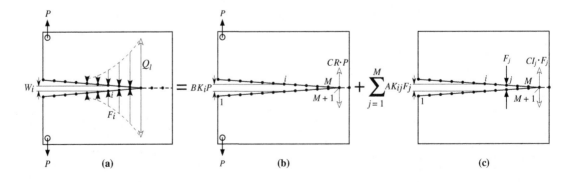

FIGURE 3.4 Decomposition of the fictitious crack problem: (a) as the superposition of a cracked structure subjected to, (b) the external load P, and (c) a set of closing nodal forces F_j.

If the external load P is regarded as an unknown in the preceding formulation, the critical condition for crack propagation must be established at the crack tip by exploiting the linear relations of the tip-nodal force with the external load and the cohesive forces of the crack (Ohtsu, 1990), as

$$Q_l = CR \cdot P + \sum_{i=1}^{M} CI_i \cdot F_i \tag{3.11}$$

where Q_l is the critical nodal force (the tensile strength of concrete times the surface area apportioned to the node at the crack tip), CR is the nodal force at the crack tip produced by a unit external load, and CI_i is the nodal force at the crack tip produced by a pair of unit closing forces applied at the ith node of the crack. Again, the coefficients CR and CI_i can be determined by FE computations. With this additional equation the propagation of a fictitious crack can now be calculated by solving $2M + 1$ equations for $2M + 1$ unknowns. Since the growth of the crack is controlled by stipulating the critical stress state at the crack tip, this method is called the crack-tip-controlled method, and the corresponding equations are called crack equations.

As seen from the preceding discussion, Petersson's formulation relies on a large number of predetermined influence coefficients with which the characteristic equations of the fictitious crack are established based on the principle of superposition. In decomposing the original problem into a set of simple elastic cases, however, a stress singularity is encountered at the crack tip when the cracked body is subjected to an external load, as shown in Figure 3.4b. In general, the stress singularity at the crack tip would render the calculation of the tip force meaningless. In the case of the fictitious crack model in concrete, however, it can be proved that this stress singularity can reasonably be ignored because it is too weak to seriously affect the accuracy of the elastic solution. A detailed analysis on this problem will be presented following.

3.3 THE PRINCIPLE OF SUPERPOSITION

The principle of superposition in elasticity states that the solution of an elastic problem with a complex loading system can be obtained by subdividing the original problem into a set of elastic problems with simpler loading conditions and by summing up the solutions of these individual problems in terms of stress, strain, and displacement. This principle can be verified as follows.

Suppose that a linear elastic body is subjected to single loads $\vec{P}_1$ and $\vec{P}_2$, respectively, as illustrated in Figure 3.5a. By solving these two problems, the obtained stresses $\sigma^1_{ij}(x, y)$ and $\sigma^2_{ij}(x, y)$ must satisfy the equilibrium conditions in Eq. (2.3). This ensures that the combined stresses $\sigma_{ij}(x, y) = \sigma^1_{ij}(x, y) + \sigma^2_{ij}(x, y)$ automatically satisfy the equilibrium conditions. Similarly, the obtained strains $\varepsilon^1_{ij}(x, y)$ and $\varepsilon^2_{ij}(x, y)$ must satisfy the compatibility condition in Eq. (2.2). Hence, the combined strains $\varepsilon_{ij}(x, y) = \varepsilon^1_{ij}(x, y) + \varepsilon^2_{ij}(x, y)$ also satisfy the compatibility condition. Apparently, the linear relationship between $\sigma^1_{ij}(x, y)$ and $\varepsilon^1_{ij}(x, y)$, and $\sigma^2_{ij}(x, y)$ and $\varepsilon^2_{ij}(x, y)$ also exists between $\sigma_{ij}(x, y)$ and $\varepsilon_{ij}(x, y)$. Thus, it can be concluded that these combined stresses and strains are the solutions of the problem in which the linear elastic body is subjected to the combined loads of $\vec{P}_1$ and $\vec{P}_2$, as shown in Figure 3.5b, provided that the boundary conditions of the problem are satisfied.

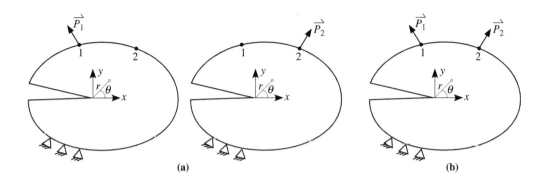

FIGURE 3.5 Elastic body with single and composite loads: (a) elastic body with single loads, and (b) elastic body with composite loads.

Notice that the principle of superposition applies to stress intensity factors, too. For example, the stress intensity factor under the composite loads in Figure 3.5b can be obtained by summing up the two stress intensity factors due to the single loads in Figure 3.5a—for example,

$$K_I = K_I^{(1)} + K_I^{(2)} \tag{3.12}$$

By virtue of this principle, stress intensity solutions for complex configurations can be obtained from simpler cases for which the solutions are well established.

3.4 THE RECIPROCITY PRINCIPLE

In Figure 3.5a, let δ_1' and δ_2' represent the displacements of the points 1 and 2 in the directions of the forces for the first condition of loading $\vec{P}_1$, and δ_1'' and δ_2'' for the second condition of loading $\vec{P}_2$. Based on the principle of superposition, the displacements in Figure 3.5b due to the simultaneous loading of $\vec{P}_1$ and $\vec{P}_2$ are given by $\delta_1' + \delta_1''$ and $\delta_2' + \delta_2''$. The total strain energy stored in the body is

$$U^{(1)} = \frac{1}{2}[(\delta_1' + \delta_1'')P_1 + (\delta_2' + \delta_2'')P_2] \tag{3.13}$$

For a specific order of loading, such as applying the load $\vec{P}_2$ first and then the load $\vec{P}_1$, the total strain energy becomes

$$U^{(2)} = \frac{1}{2}\delta_2''P_2 + \frac{1}{2}\delta_1'P_1 + \delta_2'P_2 \tag{3.14}$$

For linear elastic materials the amount of strain energy does not depend on the order in which the loads are applied. Equating the obtained strain energies leads to

$$\delta_1''P_1 = \delta_2'P_2 \tag{3.15}$$

For unit loads $P_1 = P_2 = 1$, the reciprocity principle is proved:

$$\delta_1'' = \delta_2' \tag{3.16}$$

Equation (3.16) states that the displacement at the point of the second loading in the load direction due to a unit load at the point of the first loading equals the displacement at the point of the first loading in the load direction due to a unit load at the point of the second loading.

3.5 THE SINGULARITY ISSUE

The issue of stress singularity in evaluating the tip tensile force using the FE model in Figure 3.4b, where a traction-free fictitious crack is subjected to tension loads, must be addressed (Shi, 2004). Linear-elastic stress analysis of sharp cracks predicts infinite stresses at the crack tip. In an FE computation this problem is rarely encountered because the tensile stress at the crack tip is often obtained as the averaged stress from the surrounding continuum elements. However, is the value thus obtained valid? As discussed in the preceding chapter, if the size of plastic zone is extremely small and well within the singularity-controlled region, as shown in Figure 2.9, the deviation of the stress field obtained by linear-elastic stress analysis in the vicinity of the crack tip from the actual stress field can be ignored. This is considered to be the case in the crack analysis of concrete because the fictitious crack model assumes that the FPZ is long and infinitesimally narrow (Gerstle and Xie, 1992). Consequently, the size of the plastic zone at the fictitious crack tip, where imaginary yielding is assumed to take place at the tensile strength of concrete, is extremely small. An evaluation is given following.

On the crack plane ($\theta = 0$) in Figure 3.6, the first-order estimate for the size of plastic zone, r_p^* and the COD at O, ϕ, are obtained for mode-I fracture and plane-stress conditions [Eqs. (2.80) and (2.87)], as

$$r_p^* = \frac{1}{2\pi}\left(\frac{K_I}{\sigma_{ys}}\right)^2 \tag{3.17}$$

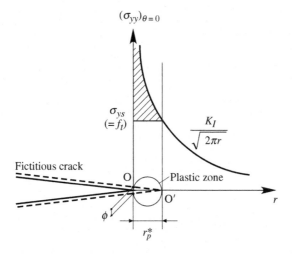

FIGURE 3.6 Imaginary inelastic crack-tip behavior in concrete and estimate of the size of the plastic zone (r_p^*).

$$\phi = \frac{4K_I^2}{\pi E \sigma_{ys}} \tag{3.18}$$

By eliminating K_I from the preceding equations and replacing the yield strength σ_{ys} with the tensile strength of concrete, f_t, the size of plastic zone is obtained as

$$r_p^* = \frac{1}{8} \frac{E}{f_t} \phi \tag{3.19}$$

As seen, with E/f_t being a material constant, r_p^* is proportional to ϕ. Because the COD is infinitesimally narrow near the tip of the fictitious crack (in numerical simulations the obtained near-the-tip COD is generally less than $W_c/100$), an extremely small r_p^* can be expected. For example, assuming $W_c = 0.1$ mm and $\phi = W_c/100 = 0.001$ mm, and $E/f_t = 6000$ for ordinary concrete, Eq. (3.19) leads to $r_p^* = 0.75$ mm. If the stress redistribution in the shaded area in Figure 3.6 is taken into account, the second-order estimate in Eq. (2.83) becomes $r_p = 2r_p^* = 1.5$ mm. Compared with the typical dimension of concrete structural members and the size of cracks, the value of several millimeters for the size of the plastic zone is indeed negligible.

3.6 CRACK PATH MODELING WITH DUAL NODES

In the discrete modeling approach, the crack surface constitutes a part of the geometric shape of the problem. Therefore, during crack analysis the boundary condition of the FE model continuously varies. In order to cope with this change of boundary condition efficiently, a path-modeling method using the so-called dummy elements and dual nodes is presented following, with which a fictitious crack can be easily introduced into the FE model along a preset crack path (Shi et al., 2001).

As shown in Figure 3.7, in an FE model a presumed crack path is modeled by pairs of dual nodes, which are individual nodes that share the same coordinates and are linked together rigidly through inner connections of extremely large spring coefficients. These dual nodes are introduced into an FE model through dummy elements, which are imaginary elements and are solely composed of dual nodes. Obviously, a dummy element is not subjected to the stress analysis. Based on this modeling technique, introducing a fictitious crack into a continuous body is as simple as setting the spring coefficients to zero to sever the connections between dual nodes. In the FE formulation, the local stiffness matrix of a dummy element, as illustrated in Figure 3.7, is expressed as

$$[K]_{dummy}^{(i)} = \begin{bmatrix} k_{11}^{(i)} & k_{12}^{(i)} & k_{13}^{(i)} \\ k_{21}^{(i)} & k_{22}^{(i)} & k_{23}^{(i)} \\ k_{31}^{(i)} & k_{32}^{(i)} & k_{33}^{(i)} \end{bmatrix} = \begin{bmatrix} k & 0 & -k \\ 0 & 0 & 0 \\ -k & 0 & k \end{bmatrix} \tag{3.20}$$

with

$$k = \begin{bmatrix} \mu_n & 0 \\ 0 & \mu_t \end{bmatrix} \tag{3.21}$$

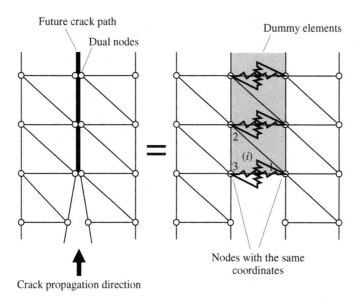

FIGURE 3.7 Crack-path modeling with dummy elements and dual nodes.

where μ_n and μ_t are the normal and the tangential spring coefficients of inner connections. Notice that these coefficients are given only two types of values: either infinity when modeling a continuum or zero when simulating a crack.

By using this path-modeling technique, the size of the problem (i.e., the total numbers of nodes and elements of the FE model) remains unchanged during numerical modeling of crack propagation. This feature is extremely important because it simplifies the programming logic for obtaining numerical solutions. This is especially so when dealing with multiple-crack problems and modeling curvilinear crack propagation that usually requires remeshing.

3.7 THE REMESHING SCHEME FOR AN ARBITRARY CRACK PATH

In general, cracks in concrete structures are curvilinear due to the existence of shear force on the crack surface, and their exact locations are unknown prior to numerical analysis, so remeshing is necessary for modeling discrete crack propagation. To minimize the computational burden, a simple scheme was developed by Shi et al. (2003) for automatic remeshing of curvilinear crack trajectories at each step of crack propagation, without altering the total numbers of nodes and meshes.

To simplify the rules for mesh reforming, the finite element meshes employed are confined to triangular elements that are connected in a regular manner. As shown in Figures 3.8 through 3.10, after obtaining a new stress field, the next-step crack extension is set in the direction normal to the tip tensile force Q, and the presumed future crack path marked by a row of dual nodes ahead of the crack is then shifted parallel by moving the nearest nodes of corresponding elements to the

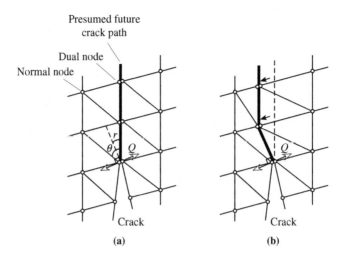

FIGURE 3.8 Remeshing scheme—(a) before and (b) after—for left-curving without interchanges of normal and dual nodes ($r \leq \theta/2$).

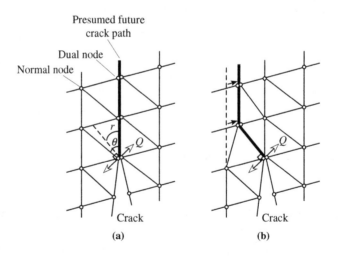

FIGURE 3.9 Remeshing scheme—(a) before and (b) after—for left-curving with interchanges of normal and dual nodes ($\theta/2 < r \leq \theta$).

new positions indicated in each case. Except for a few interchanges between the normal nodes and the dual nodes in the two cases shown in Figures 3.9 and 3.10, no additional nodes and meshes need to be generated in this process. Although Figures 3.8 through 3.10 illustrate only the cases when the crack curves to the left, the same rules apply when the crack curves to the right.

As shown in Figure 3.11, in the equivalent cases of Figures 3.9 and 3.10, reconstructions of the two meshes on the right side of the crack are required prior to constructing the new crack path.

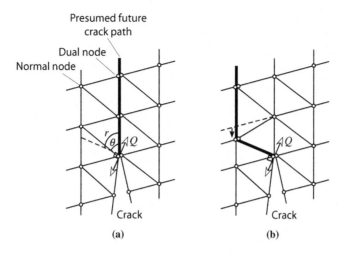

FIGURE 3.10 Remeshing scheme—(a) before and (b) after—for left-curving with interchanges of normal and dual nodes ($r > \theta$).

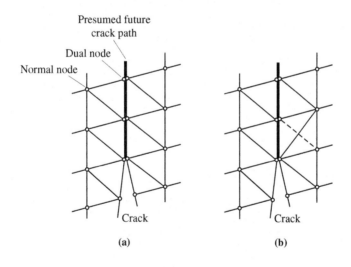

FIGURE 3.11 Remeshing scheme—(a) before and (b) after—for right-curving: changing element compositions and forming new meshes.

This simple scheme should be used with caution because it may result in ill-shaped elements. This problem can be avoided by readjusting the shapes of the modified elements of poor aspect ratio after remeshing, while keeping the newly revised future crack path unchanged. Apparently, some additional computations are required to recalculate the stress and displacement fields before moving on to the next step of the crack analysis.

3.8 THE SOLUTION SCHEME FOR INCREMENTAL STRESS ANALYSIS

Due to variations of the geometric boundary condition in any two successive steps of numerical computation, stress analysis in the discrete approach is most easily carried out in terms of the total stress and strain. This type of stress analysis has limitations because it is incapable of taking into account the dependence of inelastic behavior on loading history, especially when prominent plastic deformation takes place in the compression area of a structure. To overcome this difficulty and to make the FCM available for wider applications, a numerical technique based on the incremental procedure (Shi and Nakano, 1998; Shi et al., 2001) is introduced next.

As shown in Figure 3.12, for any given load P, the initial displacement of the structure is estimated as

$$^{m+1}u^{(1)} = {}^{m+1}u^{(0)} + ({}^{m}u - {}^{m}u^{(0)}) \tag{3.22}$$

with

$$^{m}u^{(0)} = {}^{m}K_0^{-1} \cdot {}^{m}P \tag{3.23}$$

$$^{m+1}u^{(0)} = {}^{m+1}K_0^{-1} \cdot {}^{m+1}P. \tag{3.24}$$

Here, $^{m+1}u^{(1)}$ is the initial displacement at the load level ^{m+1}P, ^{m}u is the converged solution at the previous load level ^{m}P, and $^{m}u^{(0)}$ and $^{m+1}u^{(0)}$ are the reference positions at the respective load levels. Notice that the changing of the initial stiffnesses $^{m}K_0$ and $^{m+1}K_0$ is the result of the variation of boundary conditions as cracks propagate. Thus, the gradual weakening of the structural stiffness due to cracking is monitored in the reference positions $^{m}u^{(0)}$ and $^{m+1}u^{(0)}$.

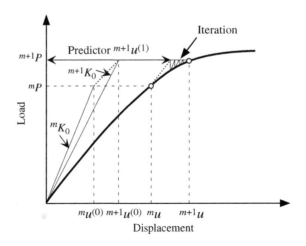

FIGURE 3.12 Numerical scheme for incremental stress analysis (after Shi et al. 2001; courtesy of ASCE).

The incremental displacement

$$\Delta u = {}^{m+1}u^{(1)} - {}^{m}u = {}^{m+1}u^{(0)} - {}^{m}u^{(0)} \tag{3.25}$$

is then used during the first iteration to modify the predictor ${}^{m+1}u^{(1)}$, and the remaining iterations follow the Newton-Raphson procedure to eliminate the residual forces until the solution converges to ${}^{m+1}u$.

Therefore, during the whole process of iterative computations, the stress analysis is carried out incrementally. The effectiveness of this numerical technique is based on the fact that in the iterative solution of a numerical problem, the convergence of the numerical solution can eventually be achieved through iterations, provided the predictor is reasonable.

REFERENCES

Bazant, Z. P., and Planas, J. (1998). *Fracture and Size Effect in Concrete and Other Quasibrittle Materials,* pp. 190–199, CRC Press.

Elices, M., and Planas, J. (1989). "Material models." In *Fracture Mechanics of Concrete Structures,* L. Elfgren ed., pp. 16–66, Chapman and Hall.

Evans, R. H., and Marathe, M. S. (1968). "Microcracking and stress-strain curves for concrete in tension." *Mater. Struct.* 1(1), 61–64.

Gerstle, W. H., and Xie, M. (1992). "FEM modeling of fictitious crack propagation in concrete." *J. Engineering Mechanics,* 118(2), 416–434.

Hillerborg, A., Modeer, M., and Petersson, P. E. (1976). "Analysis of crack formation and crack growth in concrete by means of fracture mechanics and finite elements." *Cement and Concrete Research,* 6(6), 773–782.

Hughes, B. P., and Chapman, G. P. (1966). "The complete stress-strain for concrete in direct tension." *RILEM Bulletin,* 30, 95–97.

Ohtsu, M. (1990). "Tension softening properties in numerical analysis." *Colloquium on Fracture Mechanics of Concrete Structures,* pp. 55–65, JCI Committee Report.

Petersson, P. E. (1981). *Crack Growth and Development of Fracture Process Zone in Plain Concrete and Similar Materials,* Report No. TVBM-1006. Division of Building Materials, Lund Institute of Technology, Lund, Sweden.

Planas, J., Elices, M., and Guinea, G. V. (1995). "The extended cohesive crack." In *Fracture of Brittle Disordered Materials: Concrete, Rock and Ceramics,* G. Bakker and B. L. Karihaloo eds., pp. 51–65, E&FN Spon, London.

Shi, Z. (2004). "Numerical analysis of mixed-mode fracture in concrete using extended fictitious crack model." *J. Struct. Eng.,* 130(11), 1738–1747.

Shi, Z., and Nakano, M. (1998). "Numerical approach based on the energy criterion in fracture analysis of concrete structures." *Proceedings of FRAMCOS-3,* H. Mihashi and K. Rokugo eds., pp. 1015–1024, AEDIFICATIO Publishers.

Shi, Z., Ohtsu, M., Suzuki, M., and Hibino, Y. (2001) "Numerial analysis of multiple cracks in concrete using the discrete approach." *J. Struct. Eng.,* 127(9), 1085–1091.

Shi, Z., Suzuki, M., and Nakano, M. (2003) "Numerical analysis of multiple discrete cracks in concrete dams using extended fictitious crack model." *J. Struct. Eng.,* 129(3), 324–336.

Extended Fictitious Crack Model for Multiple-Crack Analysis

4.1 INTRODUCTION

In many engineering applications involving an aging or partially damaged concrete structure, the available information on structural integrity is often limited to the crack-opening widths of several distinct cracks in the structure. Take as an example the concrete lining of an aging waterway tunnel, one of the major constituents of a hydraulic power facility. It is known that a frequent cause of cracking in waterway tunnels is the formation of voids behind the concrete lining in the ceiling area that causes structural deformation to progress under external compression, leading to the propagation of several distinct longitudinal cracks in the arch areas and the sidewalls. Obviously, for evaluating the structural safety of these aging waterway tunnels, crack-opening widths serve as an important index.

Cracking in concrete dams is another case in point. Based on circumstantial evidence drawn from documented reports of past incidents involving the excessive cracking of several concrete dams (Ingraffea, 1990; Zhang and Karihaloo, 1992; Feng et al., 1996), structurally damaging cracks in dams often start from surface cracks that are formed during construction. It is known that surface cracks often appear in multiple numbers, and depending on the circumstances, their number could vary from several cracks to several tens of them. Most of these surface cracks will later be proved inactive and will not cause serious structural problems. Hence, the real challenge for the crack analysis of concrete dams is to identify those that are potentially active and to predict their propagation paths under various possible loading conditions. Countermeasures can then be taken at an early stage to stem their further growth. What compels a seemingly ordinary surface crack to grow into a major structural crack is a complex problem involving crack interaction, and obviously its solution requires a discrete crack approach.

Following the development of the fictitious crack model (FCM) by Hillerborg and his colleagues (Hillerborg et al., 1976) for analyzing the cracking behavior of a single crack, less progress was made in extending the method to multiple-crack problems despite extensive research efforts. The core issue here is the difficulty in determining the true cracking mode for the nonlinear crack problem, which requires the most active crack or cracks to be identified among a group of potentially active cracks during each load increment in numerical analysis.

The complexity of the problem solution also stems from its computational aspect. Apparently, the numerical treatment of several varying crack surfaces requires a certain degree of flexibility and sophistication in the modeling techniques. With increasing demand for accuracy in the crack analysis of concrete structures and the maturation of computational techniques, a numerical analysis theory that extends the FCM to multiple-crack analysis emerged in the study of cracking behaviors in tunnel linings and concrete dams (Shi et al., 2001, 2003). The key point in its solution strategy is to formulate a multiple-crack problem based on various relevant cracking modes and to identify the true cracking mode by applying energy principles to the problem. The final solution is then obtained under the boundary condition that reflects the true cracking behavior.

In the following sections, the core issues in solving multiple-crack problems will be addressed, and the solution strategy will be elucidated. The numerical formulation begins with a single-crack problem, and the crack equation that leads to the solution of the unknown boundary condition is established using a crack-tip-controlled modeling method. Then the crack equations for a multiple-crack problem are formulated with the crack interactions taken into account explicitly, and the validity of the numerical solutions is discussed. As illustrative examples, crack analysis will be carried out on three types of structures or structural members: simple beams, tunnel linings, and concrete dams.

In the beam problem, the variables related to cracks include the number of initial notches, their positions, their sizes, and their orientations. The obtained cracking behaviors are examined to illustrate the effectiveness of this method in dealing with multiple discrete cracks. In the tunnel-lining problem, the failure process of a real-scale tunnel-lining specimen under compressive loads is analyzed, and the fracture behaviors obtained are compared with the experimental observations. Finally, the fracture tests on scale models of a gravity dam by Carpinteri et al. (1992) are studied, which initially contained a single notch of two sizes. By introducing multiple initial notches into the FE models, numerical modeling is extended beyond the original single-crack problem to allow multiple cracks to propagate, reproducing some typical cracking behaviors in concrete dams.

4.2 CORE ISSUES AND SOLUTION STRATEGY

As stated earlier, the Griffith energy theory (1921, 1924) is the basis for understanding material fracture. Based on this theory, an elastic body under tension loads experiences cracking if the strain energy stored in the body is sufficient to supply the fracture energy required for creating a new crack surface so as to achieve the state of minimum potential energy at a given load level. When multiple cracks are involved, the propagation of cracks can take place in various cracking modes. Figures 4.1 and 4.2 show two solid bodies containing, respectively, a single crack and two arbitrary cracks. In the case of the single crack, there is no ambiguity concerning the next-step cracking mode because under loading the crack has only one mode of reaction: to propagate. Therefore, in numerical analysis the next-step crack path, which constitutes a part of the geometric boundary condition of the problem, can be predetermined. On the other hand, each of the two cracks in Figure 4.2a possesses three possible modes of motion as the loading increases: crack propagation, crack arrest, and crack closure. Combinations of these modes between the two cracks could lead to five potential cracking modes, and in each scenario at least one crack is active, as shown in Figure 4.2b. Obviously, each of these cracking modes leads to a boundary value problem that should be solved individually. Identification of the true cracking mode requires further analysis on the validity of the obtained numerical solutions and a consideration of the energy principles.

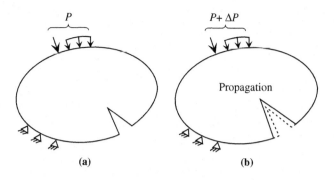

FIGURE 4.1 Cracking modes for single cracks: (a) before incremental loading and (b) after incremental loading.

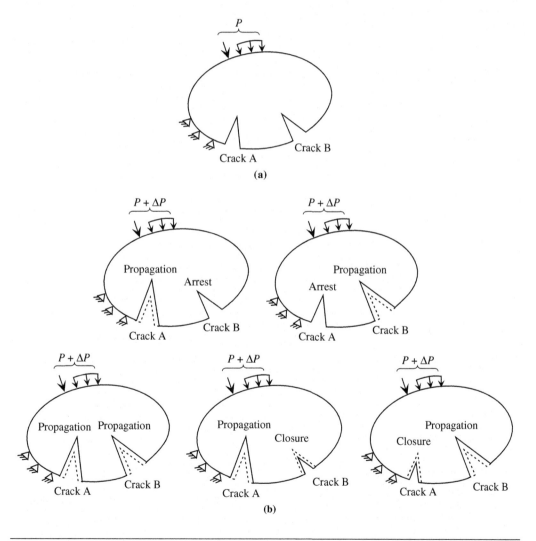

FIGURE 4.2 Cracking modes for multiple cracks: (a) before incremental loading and (b) after incremental loading.

As our vision on how to solve a multiple-crack problem using the FCM becomes clearer, so is the difficulty that lies ahead. Since each crack has three potential modes of motion, combinations of these modes among all the cracks concerned give rise to the possible crack propagation patterns or cracking modes. For an arbitrary number of cracks, N, the number of cracking modes, NCM, is readily obtained as

$$NCM = 3^n - 2^n \tag{4.1}$$

It should be noted that inactive cracking modes that contain only the modes of crack arrest and crack closure are irrelevant and thus must be eliminated from Eq. (4.1). Even with only five cracks, the NCM well exceeds 200, and with ten, it approaches 60,000. Obviously, a direct search for the true cracking mode from all the possible cracking modes given by Eq. (4.1) is out of the question. Such an approach is not analytic, being neither computationally feasible nor logistically sound. In general, for any specific problems only some of these cracking modes could lead to geometrically admissible strain fields and thus valid solutions. The relevance of a cracking mode to a specific problem depends on the problem itself, including the structural geometry and loading conditions, the locations of the cracks, as well as the crack interactions. The primary obstacle in applying the FCM to solve multiple-crack problems is the difficulty of ascertaining the true cracking mode from all of the possibilities given by Eq. (4.1), each time the load increases.

To overcome this difficulty, an effective solution scheme needs to be devised. This solution strategy should be flexible enough to allow all of the relevant cracking modes to be considered and at the same time be computationally feasible. In the extended fictitious crack model (EFCM) as proposed by Shi et al. (2001), a solution strategy based on the single-active-crack modes (see the first two modes in Figure 4.2b) is adopted. In solving the problem, each crack is individually assumed to be an active crack, and the basic equations governing the growth of this single crack are derived. The external load required for propagating this particular crack is calculated by solving crack equations. Among all the solutions obtained, the true active crack is then determined based on a minimum load criterion that stipulates the propagation of an active crack at the minimum load. Since this criterion effectively ensures that the body achieves the state of minimum potential energy by transferring the body's strain energy into fracture energy in the process of propagating the true active crack, it is in essence equivalent to an energy criterion.

It should be noted that when an assumed cracking mode is irrelevant to the problem, invalid solutions with geometrically inadmissible strain fields could be encountered. By resetting the tips of certain restrained cracks based on the feedback from these erroneous solutions and recalculating the case, a variety of cracking modes may emerge, which include simultaneous crack propagations of several cracks and crack growth accompanied by crack closure (see the last three modes in Figure 4.2b). As the true cracking mode is found, the stress and displacement fields can now be calculated under the relevant boundary condition, which includes the true crack paths for every crack and the cohesive forces (obtained by solving the crack equations) acting on these cracks, as well as the minimum external loads.

This solution strategy is justifiable because, among all the possible cracking modes of a given crack, the single-active-crack mode requires the least amount of fracture energy for the crack to propagate. Any other modes involving simultaneous propagations or closures of other cracks inevitably require extra external energy to extend or close these cracks (as closure of a crack also requires external energy). Hence, it is logical to start with the single-active-crack mode of each crack in a crack analysis. If such a mode is proved to be erroneous or unrealistic under the specific boundary condition of a problem (implying strong crack interactions), it will eventually be replaced by other modes as a result of readjusting the tip positions of certain cracks. The single-active-crack mode approach is also consistent with the fact that in the initial stage of crack propagation when all the cracks are small and the crack interactions are insignificant, the growth of each crack is practically independent of all the others.

The validity of this analytical theory in accurately predicting cracking behavior in a multiple-crack situation has been verified through the numerical analysis of a large number of cracked tunnel-lining problems that include both waterway tunnels and highway tunnels (some of these cases will be discussed in Chapter 8). Experimental verifications with bending tests of plain concrete beams that contain multiple initial notches have also been very successful in proving the accuracy of numerical predictions on the failure modes of these notched beams, which will be discussed in Chapter 5 and Chapter 6.

4.3 NUMERICAL FORMULATION OF A SINGLE-CRACK PROBLEM

We now examine the analytical concept of the crack-tip-controlled modeling of a single crack, formulated and referred to as the crack equations by Ohtsu (1990). The method is chosen because it offers an easy way for regulating cracking behavior, which will be very useful when modeling multiple cracks. Figure 4.3 shows a single crack of the mode-I type, propagating in the direction normal to the maximum principal tensile stress at the tip of the crack. The basic equations governing the next-step crack growth (by moving the present tip of the crack to the next node along the future crack path) are obtained using the principle of superposition. Assuming the solid body to be linear elastic, the overall equations are derived by linear combinations of individual solutions that use the influence coefficients due to the external load (Figure 4.3a) and the cohesive forces at the fictitious crack (Figure 4.3b), respectively. Notice that the external load and the cohesive forces are unknowns and will be determined by solving the following crack equations.

As shown in Figure 4.3, the crack and its fictitious prolongation are represented by separating pairs of dual nodes, a special kind of node consisting of two nodal points of the same coordinates that are rigidly connected to each other before separation. In formulating crack equations, subscript l represents the limit value of a nodal force, and superscripts i and j denote the ith and jth nodes. Let CR represent the reaction at the tip of the crack under a unit external load, and CI^i denote the tip force due to a pair of unit cohesive forces at the ith node of the fictitious crack. For the crack to propagate, the tip tensile force must reach the tensile strength of concrete, given by

$$Q_l = CR \cdot P + \sum_{i=1}^{M} CI^i F^i \tag{4.2}$$

where Q_l is the limit nodal force (the tensile strength of concrete times the surface area apportioned to a nodal point); P is the external load; F^i represents the pair of cohesive forces at the ith node; and M is the number of nodes inside the fictitious crack. Notice that Eq. (4.2) is the basic equation for stable crack propagation.

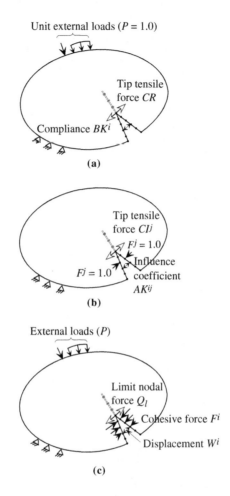

FIGURE 4.3 Concept of the crack-tip-controlled modeling of single cracks: (a) forces and displacements at the crack due to unit external loads, (b) forces and displacements at the crack due to a pair of unit cohesive forces, and (c) load condition for crack propagation.

The crack-opening displacement (COD) at the ith node is given by

$$W^i = BK^i \cdot P + \sum_{j=1}^{M} AK^{ij} F^j \tag{4.3}$$

where $i = 1, \ldots, M$. Here, BK^i is the compliance at the ith node due to the external load P. The influence coefficient AK^{ij} is the COD at the ith node due to a pair of unit cohesive forces at the jth node (Figure 4.3b). By virtue of the reciprocity theorem, $AK^{ij} = AK^{ji}$.

The geometric shapes of the fictitious crack are determined by Eq. (4.3). Notice that CR, CI^i, BK^i, and AK^{ij} are all obtained by linear elastic FE computations based on the FE models shown in Figures 4.3a and b.

Along the fictitious crack, the cohesive forces F^i and the crack-opening displacements W^i must obey the tension-softening law of concrete

$$F^i = f(W^i) \tag{4.4}$$

where $i = 1, \ldots, M$.

Equations (4.2) to (4.4) form the so-called crack equations, which are composed of the propagation condition, the shape function, and the stress-COD relations of the fictitious crack. These equations represent a mathematical formulation of the crack propagation in a single-crack problem. With the number of equations $(2M + 1)$ matching the number of unknowns $(2M + 1)$, the problem can be uniquely solved to obtain the external load and the cohesive forces, since these equations are linearly independent. Then the stress and displacement fields with the next-step crack propagation can be calculated under the newly obtained boundary condition, as shown in Figure 4.3c. This computational process will be continued until structural failure. As such, the crack equations provide the theoretical basis for crack analysis, which is a continuing process of finding the loading history and the equilibrium states for a propagating crack. The basic solution procedure is illustrated in Figure 4.4.

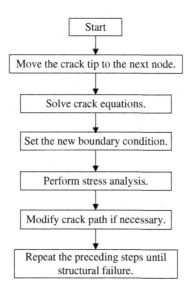

FIGURE 4.4 Solution procedure for the crack-tip-controlled modeling of single cracks.

4.4 NUMERICAL FORMULATION OF A MULTIPLE-CRACK PROBLEM

The fundamental difference between single-crack and multiple-crack problems is whether or not the next-step cracking behavior is predictable. The mathematical formulation for single-crack problems discussed in the previous section is based on the mode of crack propagation, which is the only valid mode for a single crack under tension. When multiple cracks are involved, the next-step cracking behavior cannot be uniquely determined. Therefore, crack equations for a multiple-crack problem should be established by considering multiple cracking modes. The following formulations are based on the single-active-crack modes discussed previously.

Figure 4.5 illustrates two cracks of the mode-I type, crack A and crack B, where crack propagation is set in the direction normal to the tensile force at the tip of each fictitious crack. In formulating crack equations, subscripts a and b represent, respectively, crack A and crack B, and l stands for the limit value of a nodal force. Superscripts i, j, and k denote the corresponding nodes at designated cracks. For clarity, the cohesive forces and the CODs of the inactive crack are marked by asterisks. To begin with, crack A is assumed to be the sole propagating crack, so the tensile force at its tip must reach the nodal force limit Q_{la}, given by

$$Q_{la} = CR_a \cdot P_a + \sum_{i=1}^{N} CI_{aa}^i F_a^i + \sum_{j=1}^{M} CI_{ab}^j F_b^{*j} \tag{4.5}$$

where N and M are the number of nodes inside each fictitious crack, respectively. Notice that the tensile forces at the tip of crack A—CR_a, CI_{aa}^i, and CI_{ab}^j—are due to a unit external load, a pair of unit cohesive forces at the ith node of crack A, and a pair of unit cohesive forces at the jth node of crack B, respectively. The external load P_a is the required load for propagating crack A, while crack B remains inactive. It should be noted that the tip force components due to the cohesive forces of crack B in Eq. (4.5) represent exactly the effect of crack interaction on crack A.

The CODs along the two fictitious cracks are given by

$$W_a^i = BK_a^i \cdot P_a + \sum_{k=1}^{N} AK_{aa}^{ik} F_a^k + \sum_{j=1}^{M} AK_{ab}^{ij} F_b^{*j} \tag{4.6}$$

$$W_b^{*j} = BK_b^j \cdot P_a + \sum_{i=1}^{N} AK_{ba}^{ji} F_a^i + \sum_{k=1}^{M} AK_{bb}^{jk} F_b^{*k} \tag{4.7}$$

where $i = 1, \ldots, N$; $j = 1, \ldots, M$. Here, the compliances BK_a^i at crack A and BK_b^j at crack B are due to the external load. The influence coefficients AK_{aa}^{ik} and AK_{ab}^{ij} are the CODs at the ith node of crack A due to a pair of unit cohesive forces at the kth node of crack A and a pair of unit cohesive forces at the jth node of crack B, respectively. Similarly, the influence coefficients AK_{ba}^{ji} and AK_{bb}^{jk} represent the CODs at the jth node of crack B due to a pair of unit cohesive forces at the ith node of crack A and a pair of unit cohesive forces at the kth node of crack B, respectively. According to the reciprocity theorem, $AK_{aa}^{ik} = AK_{aa}^{ki}$, $AK_{bb}^{jk} = AK_{bb}^{kj}$, and $AK_{ab}^{ij} = AK_{ba}^{ji}$.

Finally, imposing the tension-softening law of concrete along each fictitious crack leads to

$$F_a^i = f(W_a^i) \tag{4.8}$$

$$F_b^{*j} = f(W_b^{*j}) \tag{4.9}$$

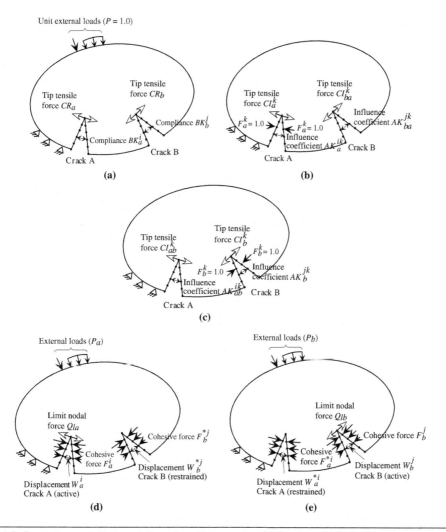

FIGURE 4.5 Crack-tip-controlled modeling of multiple cracks: (a) forces and displacements at the cracks due to unit external loads, (b) forces and displacements at the cracks due to a pair of unit cohesive forces at crack A, (c) forces and displacements at the cracks due to a pair of unit cohesive forces at crack B, (d) load condition for the growth of crack A, and (e) load condition for the growth of crack B.

where $i = 1, \ldots, N$; $j = 1, \ldots, M$. Equations (4.5) to (4.9) form the crack equations, stipulating the conditions for crack A to propagate. With the number of equations $(2N + 2M + 1)$ matching the number of unknowns $(2N + 2M + 1)$, the problem is solved uniquely to obtain the external load P_a, the cohesive forces, and the CODs at the two cracks.

Alternatively, when crack B is assumed to be the only active crack, the crack equations are derived as follows

$$Q_{lb} = CR_b \cdot P_b + \sum_{i=1}^{N} CI_{ba}^{i} F_a^{*i} + \sum_{j=1}^{M} CI_{bb}^{j} F_b^{j} \tag{4.10}$$

$$W_a^{*i} = BK_a^i \cdot P_b + \sum_{k=1}^{N} AK_{aa}^{ik} F_a^{*k} + \sum_{j=1}^{M} AK_{ab}^{ij} F_b^{j} \tag{4.11}$$

$$W_b^{j} = BK_b^j \cdot P_b + \sum_{i=1}^{N} AK_{ba}^{ji} F_a^{*i} + \sum_{k=1}^{M} AK_{bb}^{jk} F_b^{k} \tag{4.12}$$

$$F_a^{*i} = f(W_a^{*i}) \tag{4.13}$$

$$F_b^{j} = f(W_b^{j}) \tag{4.14}$$

where the external load P_b is the required load for activating crack B, while crack A is assumed to be inactive. Equations (4.10) to (4.14) form the crack equations that set the conditions for crack B to propagate. Notice that the tip force components caused by the cohesive forces of crack A in Eq. (4.10) represent explicitly the effect of crack interaction on crack B. Solving the crack equations, the external load P_b, the cohesive forces, and the CODs at the two cracks are obtained.

The equations from (4.5) to (4.14) form the two sets of crack equations required for modeling two discrete cracks, which contain the propagation condition, the shape functions, and the stress-COD relations. The influence coefficients employed in the crack equations are determined by linear elastic FE computations based on the FE models shown in Figures 4.5a–c. Upon solving the two sets of crack equations, the true cracking mode is identified based on the minimum load criterion, which predicts the onset of crack propagation at the minimum load—that is,

$$P = \min(P_a, P_b) \tag{4.15}$$

After setting the true crack paths for the next-step crack propagation, the stress and displacement fields are calculated under the condition of the obtained load and corresponding cohesive forces, as shown in Figures 4.5d–e. This process can be repeated until structural failure. The concept of the minimum load criterion for identifying the true cracking mode in crack analysis is illustrated in Figure 4.6. Obviously, the preceding solution procedure can be readily extended

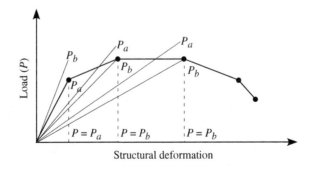

FIGURE 4.6 Concept of the minimum load criterion for crack analysis.

to problems with an arbitrary number of cracks. The flowchart of the computational procedure is shown in Figure 4.7. As seen, numerical results are checked to eliminate invalid solutions upon solving the crack equations and obtaining the stress field. These invalid solutions are encountered when an assumed cracking mode is irrelevant to the problem, and are manifested either by the tip tensile stress exceeding the tensile strength at the tip of an assumed inactive crack, or by the overlapping of the crack surfaces with negative CODs obtained at the restrained crack. In a situation like this, the crack tip is readjusted by releasing or closing the tip nodes and the problem is recalculated, as illustrated in Figure 4.7.

To close the tip of a crack, the two disconnected nodes next to the tip of the crack are reconnected, while the previous normal traction acting at these nodes is referred to as the transient

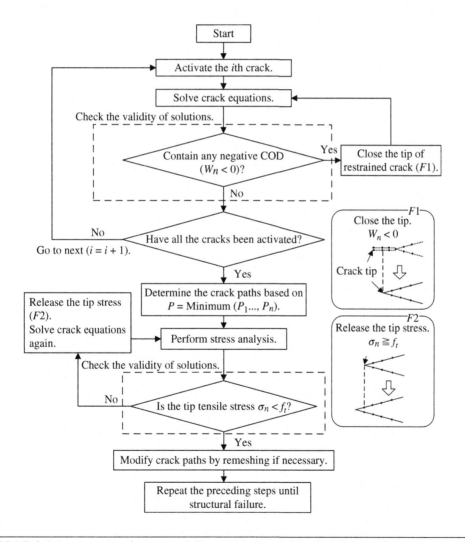

FIGURE 4.7 Solution procedure for the crack-tip-controlled modeling of multiple cracks.

material strength of the cracked material (based on the fact that the material damage due to cracking is irreversible). By readjusting the crack-tip position, other cracking modes with geometrically admissible strain fields may emerge, which include simultaneous propagations of several cracks and crack growth accompanied by crack closure. The feedback from the obtained erroneous solutions provides sufficient clues for modifying the cracking modes, and this mode-correcting function is essential for the success of the proposed computational theory in numerical analysis of multiple cracks.

In solving multiple-crack problems, the crack analysis for the next-step crack growths usually leads to multiple solutions that correspond to different load levels. The minimum load criterion stipulates that crack propagation begins at the lowest load level. This is the basis for identifying the true cracking mode from among the many potential cracking modes as defined in Eq. (4.1). Once the true cracking mode is found, a nonlinear crack problem with unknown boundary conditions is then reduced to an ordinary boundary value problem with a clearly defined boundary condition, which is solved by the established numerical procedure.

The theoretical basis for the minimum load criterion can be found in the Griffith energy principle, which states that crack extension occurs when the energy available for crack growth is sufficient to overcome the resistance of the material. Obviously, the Griffith energy principle is satisfied at the minimum load for crack propagation. Hence, the minimum load criterion is equivalent to an energy criterion for the growths of multiple cracks.

4.5 CRACK ANALYSIS OF A SIMPLE BEAM UNDER BENDING

This section presents crack analysis of a simple beam under bending with a fixed crack path that requires no path modification and with a curvilinear crack path that demands path modification—that is, remeshing.

4.5.1 Crack Analysis with a Fixed Crack Path

The first structural problem to be studied is the four-point bending tests of simple plain concrete beams (Uchida et al., 1993). An unnotched test specimen and six FE models with small initial notches are illustrated in Figure 4.8. The first two cases—Case 1-1 and Case 1-2—are half-models of the simple beam, having two initial notches each. As seen, notch A is at the midspan, and notch B is below the loading point. While for Case 1-1, notches A and B are assumed to be the same size of 10 mm, notch B in Case 1-2 is 20 mm, twice the size of the central notch. By adding one more notch next to notch B, three half-model cases with three initial notches are assumed.

As shown, the three notches in Case 2-1 are the same size—10 mm—while notch B of Case 2-2 and notch C of Case 2-3 are 20 mm, twice the size of the others. The last case, Case 2-4, is a full model of the simple beam, having three initial notches with arbitrary positions, inclinations, and sizes. Notice that the size of a small initial notch here is assumed to be 10 mm, which is one-twentieth of the beam height. For the geometric details of each case, refer to Figure 4.8.

In solving the crack equations, the bilinear tension-softening relation, as shown in Figure 4.9, is assumed (Rokugo et al., 1989). The material properties of the test specimen are summarized in Table 4.1, which include the elastic modulus E, the tensile strength f_t, the compressive strength f_c,

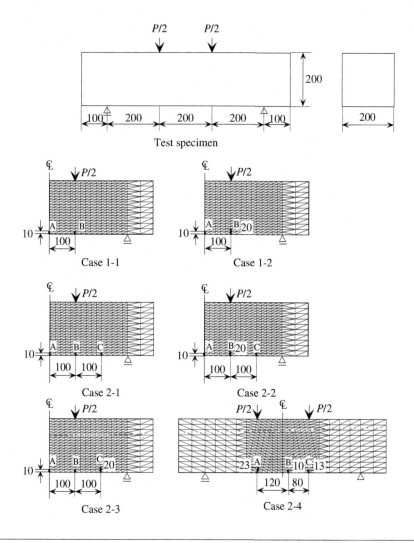

FIGURE 4.8 Fracture test of simple beam and FE models (dimensions in mm) with initial notches.

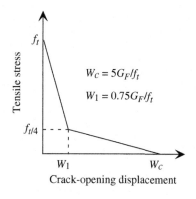

FIGURE 4.9 Bilinear tension-softening relation of concrete (Rokugo et al., 1989).

Table 4.1 Material Properties of a Simple Beam

E (GPa)	ν	f_c (MPa)	f_t (MPa)	G_F (N/mm)
27.50	0.20	33.00	2.80	0.10

Poisson's ratio ν, and the fracture energy G_F. Crack analysis will be carried out for the six numerical cases by fixing the crack paths as straight.

Two Initial Notches

The numerical results are shown in Figure 4.10, which include the load-midspan displacement relations, the load versus crack-mouth-opening displacement (CMOD) relations, and the crack

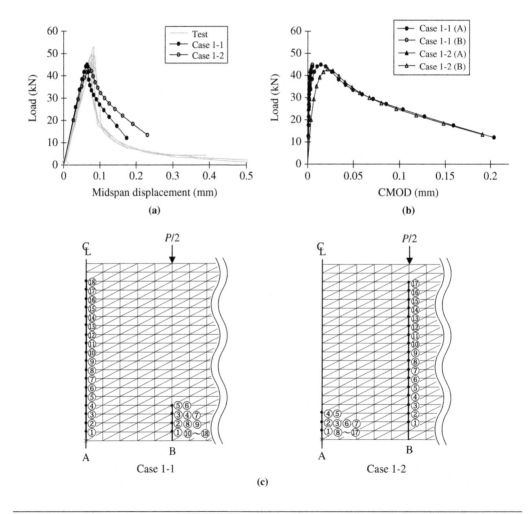

FIGURE 4.10 The numerical results of Case 1-1 and Case 1-2 with two notches: (a) load-displacement relation, (b) load-CMOD relation, and (c) crack propagation chart (after Shi et al., 2001; courtesy of ASCE).

propagation charts of the two individual cracks (where the circled number represents the tip position of that particular crack at the designated step of the crack-tip-controlled computation). With the same small size assumed for the two notches in Case 1-1, an active crack grows from the midspan notch and becomes the dominant crack, penetrating the beam almost to the end before the numerical solutions diverge at the nineteenth step.

As shown in the crack propagation chart, crack B propagates only a short distance into the beam and eventually closes after the sixth step. In Case 1-2, with a size of 20 mm assumed for notch B, the cracking behaviors are reversed: Crack B turns out to be the dominant crack, while crack A quickly becomes inactive in the early stage of the numerical computations. The obtained results clearly demonstrate how the locations and sizes of initial notches could strongly affect cracking behavior. From a purely computational point of view, to extend the inactive cracks beyond their present depths (by evoking other cracking modes) requires extra external energy— that is, a larger external load.

In the end, this may lead to a superficially large load-carrying capacity of the simple beam. Obviously, this scenario is unrealistic and potentially dangerous (as it may lead to an unsafe design) and therefore must be eliminated by implementing the minimum load criterion for crack extension. As clearly shown, the crack propagation can be traced until the very late stage in the postpeak regions, with the CMOD of each active crack reaching roughly 0.2 mm for this simple beam problem.

A comment on the obtained load-midspan displacement relations should be made. Compared with the test results, the obtained curves in the postpeak regions are less brittle than the experimental curves. This superficial stiffness in the structural response is caused by the adoption of the straight crack paths and the employment of the half-models in the numerical analysis; both are known to result in stiffer structural response. Further explanations will be given later.

Three Initial Notches

Numerical results are presented in Figure 4.11. For Cases 2-1, 2-2, and 2-3, the active cracks are crack A, crack B, and crack A, respectively. Under the given conditions, no crack emerges from notch C in the first two cases at all. A comparison between Case 2-1 and Case 1-1 reveals that the cracking behaviors in the two cases are exactly the same, indicating that notch C has no influence on crack A and crack B in Case 2-1. The same is true for Case 2-2 when compared with Case 1-2. Enlarging notch C to 20 mm in Case 2-3, a small crack appears from that notch, but it then becomes a nonpropagating crack and soon closes after the closure of crack B, as crack A penetrates deeply into the beam. The appearance of crack C in Case 2-3 slightly affects the growth of crack B due to the crack interaction, which becomes obvious when compared with Case 2-1. For clarity, only the CMODs of the active cracks are shown in the load-CMOD relations.

A comparison between Case 2-1 and Case 2-3 shows similar cracking behaviors in the two cases. This explains why the load-midspan displacement relations of the two cases are almost identical. Finally, the full-model case, Case 2-4, is analyzed to demonstrate the flexibility of the EFCM in dealing with arbitrary cracks. In this case, a main crack propagates from notch A along an inclined crack path, and the obtained structural deformation is found to be very close to the experimental curves in the postpeak regions. For details, refer to Figure 4.11.

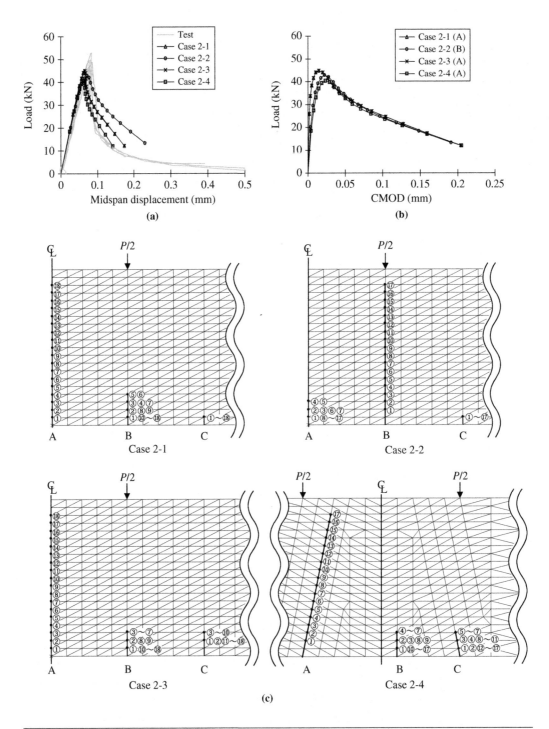

FIGURE 4.11 The numerical results of Case 2-1 to Case 2-4 with three notches: (a) load-displacement relation, (b) load-CMOD relation, and (c) crack propagation chart (after Shi et al., 2001; courtesy of ASCE).

4.5.2 **Crack Analysis with a Curvilinear Crack Path**

The fracture process of the simple beam will be restudied to simulate curvilinear crack propagation, using the simple remeshing method introduced in Chapter 3 to modify the crack paths. The unnotched test specimen and four numerical cases with small initial notches are illustrated in Figure 4.12. Cases 3-1 and 3-2 are half-models, having three initial notches each. Whereas for Case 3-1, notch B is placed under the loading point with a notch that is twice the size of notches A and C, notch B of Case 3-2 is set 4 cm away from that position to the right and has a notch size three times the others. The full-model cases—Cases 4-1 and 4-2—also contain three initial notches each. As shown in Figure 4.12, notches A, B, and C in these two cases have exactly the same random locations. The only difference is that the size of notch C in Case 4-2 is slightly smaller than its counterpart in Case 4-1.

Figures 4.13 and 4.14 present the numerical results for the half-model cases and the full-model cases, respectively, which include the load-displacement relations, the load-CMOD relations, and the crack propagation charts. Although crack propagations in curvilinear crack paths are known to involve tangential components of the cohesive forces on the crack surface, the crack analysis here is based on the mode-I assumption. In other words, the shear force at the crack surface is ignored (mixed-mode crack problems will be discussed in Chapter 7). As far as the load-displacement relations are concerned, the agreement between the experiments and the numerical analyses is deemed good enough to justify the validity of the following discussion on curvilinear crack propagations.

For the half-model cases, if all of the three initial notches in Case 3-1 were assigned equal sizes, crack A would then become the most active crack, propagating along the central line of the beam as demonstrated in Case 2-1. That case is not the focus of the present study because the crack path of the central crack is inherently vertical. To propagate a curvilinear crack, notch B of Case 3-1 is assigned a slightly larger size, and consequently crack B becomes the dominant crack, as shown in Figure 4.13. The curvilinear trajectory of crack B may seem a little puzzling, as intuition would suggest that the crack propagates vertically toward the loading point. However, with the existence of a compression field in the upper part of the beam, the crack does extend along the curvilinear trajectory as the directions of principal tensile stresses at the crack tip gradually deviate from the original horizontal orientation.

In contrast, the curvilinear trajectory of crack B in Case 3-2 is fully anticipated, since the principal tensile stress departs from the horizontal orientation from the very beginning due to the shear. It is interesting to note that, in this case, the crack tip curves toward the loading position. In both cases, the growth of crack A is limited to only one or two nodal intervals, and no crack emerges from notch C at all. The details on crack extension can be found in the propagation charts in Figure 4.13. Due to the employment of half-models in these two cases, the load-midspan displacement relations in the postpeak regions appear stiffer than the actual structural behavior.

As shown in Figure 4.14, in Case 4-1, crack A and crack C compete for simultaneous propagation up to the peak load at the sixth step, reaching almost one-half of the beam depth. Crack C becomes the dominant crack in the postpeak region, forcing the others to close eventually. Reducing the initial size of notch C by only one centimeter, however, crack A of Case 4-2 easily becomes the most progressive crack, curbing the growth of crack B and crack C at an early stage. As such, a small change in the sizes or positions of initial notches may completely alter the

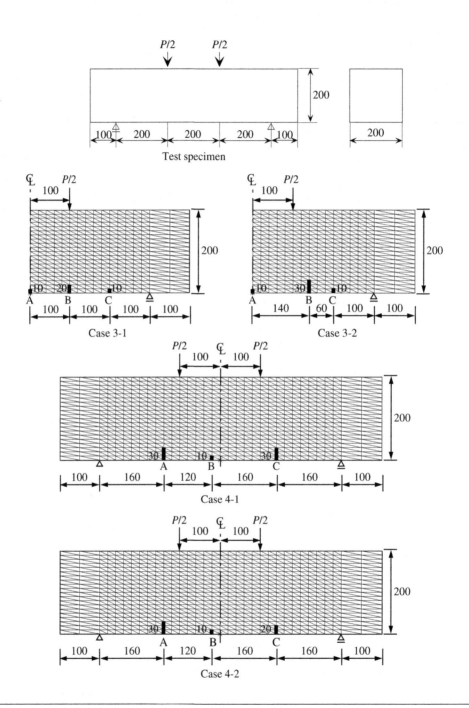

FIGURE 4.12 Fracture test of a simple beam and FE models (dimensions in mm) with initial notches.

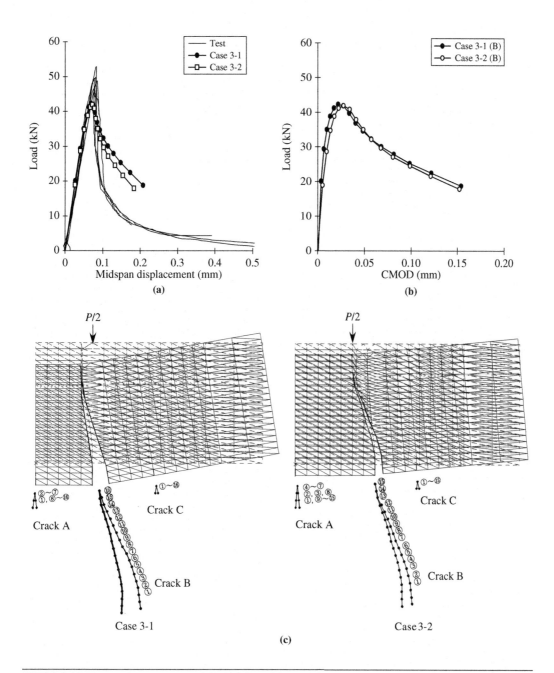

FIGURE 4.13 The numerical results of Case 3-1 and Case 3-2 with three notches: (a) load-displacement relation, (b) load-CMOD relation, and (c) crack propagation chart (after Shi et al., 2003; courtesy of ASCE).

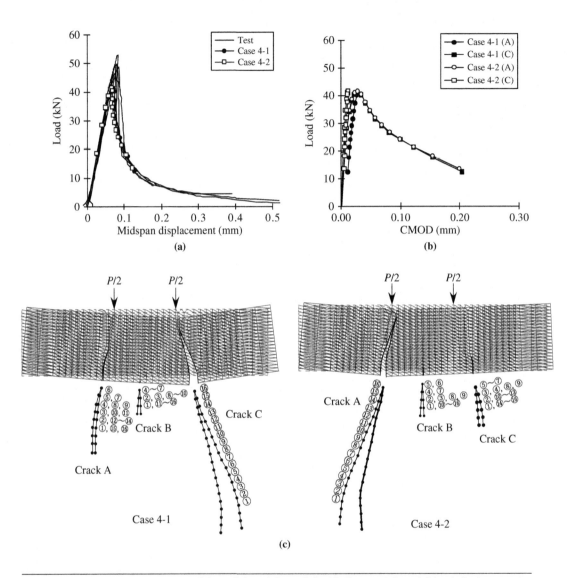

FIGURE 4.14 The numerical results of Case 4-1 and Case 4-2 with three notches: (a) load-displacement relation, (b) load-CMOD relation, and (c) crack propagation chart (after Shi et al., 2003; courtesy of ASCE).

pattern of crack propagation. This makes the prediction of cracking behavior in a real structure very difficult when multiple cracks are involved. Obviously, the crack interactions, which will be the focus of study in Chapter 5, play an important role in causing these complicated cracking behaviors. It should be pointed out that by modifying crack paths and employing full FE models, the obtained structural deformation curves almost perfectly match the experimental curves in the postpeak regions.

As demonstrated through all the beam cases studied, although multiple cracks were involved in the initial stage of crack propagation, only the dominant crack in each case became an open crack, while the others stopped propagating and closed eventually before reaching the peak load. This analytical conclusion agrees with the results of a large number of bending tests on plain concrete beams, which have invariantly shown the beam failure to be the result of a single fracture. This explains why the obtained structural response is stiffer when a half-FE-model is employed, as just discussed. By using a half-model to study the beam fracture, the obtained single open crack may imply the existence of two open cracks in the beam. In general, this failure mode is unrealistic for plain concrete beams, and it is known that unrealistic cracking modes can lead to superficially higher peak loads and stiffer structural responses, as indicated in Figure 4.6.

4.6 CRACK ANALYSIS OF A FRACTURE TEST OF A REAL-SIZE TUNNEL-LINING SPECIMEN

This section presents a fracture test of a real-size tunnel-lining specimen and the crack analysis of the test using a half-FE-model and a full-FE-model.

4.6.1 Fracture Test on a Tunnel-Lining Specimen

The second structural problem involving multiple cracks is the fracture test of a real-size concrete lining specimen of a waterway tunnel, as shown in Figure 4.15 (Abo et al., 2000). The test was carried out to investigate the cracking behavior and fracture process of a tunnel lining with void formation above the ceiling area and uniformly distributed loads applied to the sidewalls and to study the remaining load-carrying capacity after the formation of cracks. As shown, the tunnel was 2.5 m in diameter, with a wall thickness of 25 cm. The test was carried out under the load control, and during the test structural deformations were recorded at the crown and the two sidewalls along with the strains, as illustrated in Figure 4.15.

Although no measurements of the CMODs were taken during the test, the crack propagation patterns were carefully recorded, as shown in Figure 4.16, in which the load-displacement relations are also presented. As shown by the test results, five cracks propagated in the test specimen before the collapse of the tunnel under compression. The most active crack occurred in the right wall, which was followed by two progressive cracks in the bottom plate from outside. The crack in the left wall and the crack in the ceiling area from outside were small and less active. Upon reaching the peak load, the tunnel specimen failed in a brittle fashion, as indicated in the load-displacement relations. It was reported that during the experiment a certain degree of eccentric loading occurred, generating a higher pressure-load on the right wall. This is evident from the load versus displacement relations separately measured on the left and right walls, as well as from the unsymmetric crack propagation patterns recorded on the test specimen.

Based on these facts, the right portion of the bottom plate was believed to be under bending during the experiment, and the corresponding vertical supports in effect did not function, as indicated in Figure 4.16. In the following, the cracking processes of the tunnel specimen will be studied by employing two FE models: a half-model with three initial notches and a full-model with

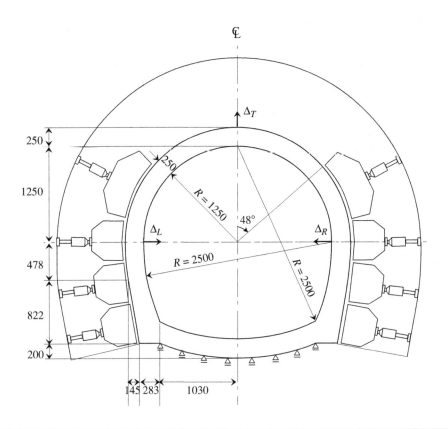

FIGURE 4.15 Fracture test (dimensions in mm) on a tunnel specimen.

five initial notches. These analyses will reveal the unique cracking behaviors frequently observed in actual tunnels. The material properties of the test specimen are summarized in Table 4.2. Notice that the bilinear tension-softening relation in Figure 4.9 is employed to solve the crack equations.

4.6.2 Crack Analysis with a Half-FE-Model

A half-FE-model is employed to study cracking behaviors in the tunnel specimen, taking into account the assumed symmetric conditions of the experiment, as shown in Figure 4.17. As seen, three initial notches are introduced into the specimen: Notch A is in the ceiling from outside, notch B is in the sidewall, and notch C is in the bottom plate from outside, next to the sidewall. The crack paths are assumed to be perpendicular to the lining surface. The locations of the notches are roughly determined from the observed crack propagation trajectories of the test results. The notch sizes are assumed to be the same at 25 mm, which is one-tenth of the wall thickness.

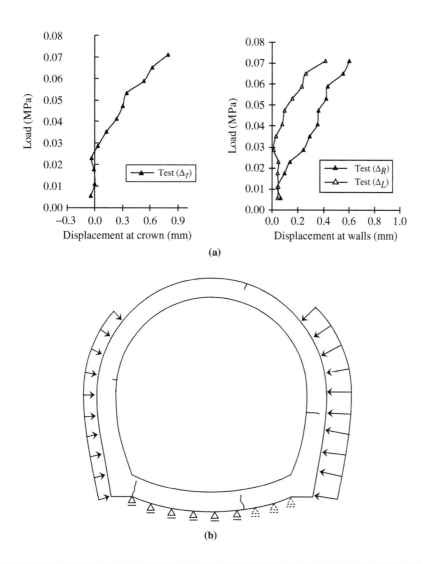

FIGURE 4.16 The results of a fracture test on a tunnel specimen: (a) load-displacement relation and (b) crack propagation chart.

Table 4.2 Material Properties of a Tunnel Specimen

E (GPa)	ν	f_c (MPa)	f_t (MPa)	G_F (N/mm)
20.00	0.20	20.00	2.00	0.10

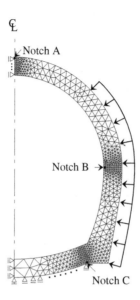

FIGURE 4.17 A half-FE-model of a tunnel specimen.

Numerical results are presented in Figure 4.18, which include the load-displacement relations at the ceiling and the sidewalls, the load-CMOD relations, and the crack propagation charts. Available experimental results are also shown. Despite some discrepancies in the loading conditions between the actual test and the numerical study, judging from the load-displacement relations it is still reasonable to conclude that the present numerical model reproduces well the general structural response of the tunnel specimen. Based on the crack propagation charts and the load-CMOD relations, crack B in the middle of the wall is found to be the most progressive crack. Crack C in the bottom plate is also active, although its growth is temporarily interrupted from the fourth to the sixth steps by the aggressive propagation of crack B. Compared with the active propagations of crack B and crack C, the slow opening of crack A in the ceiling area is quite noticeable.

As the maximum load is reached at the fourth step of the computation, crack A stops propagating, retreats, and remains inactive until structural failure. It should be pointed out that in the present case there are two propagating cracks in the postpeak regions, contrasting to the single-cracking mode in the postpeak regions of plain concrete beams. Obviously, this is due to the existence of multiple tension zones in tunnel structures. The maximum CMOD of crack B reaches 0.2 mm, and the maximum structural deformations at the ceiling and sidewalls are approximately 1 mm. The obtained cracking behaviors are clearly attributed to the existence of the void in the ceiling area, which allows the tunnel specimen to deform in the vertical direction without any restraint, thus facilitating the opening of the cracks in the sidewalls.

These numerical results highlight the importance of backfilling the voids in tunnels to stop the further opening of the cracks in tunnel linings in real engineering situations. Taking into consideration the actual eccentric load conditions in the test, the general cracking behaviors obtained from the numerical analyses are considered to be in good agreement with the experimental observations.

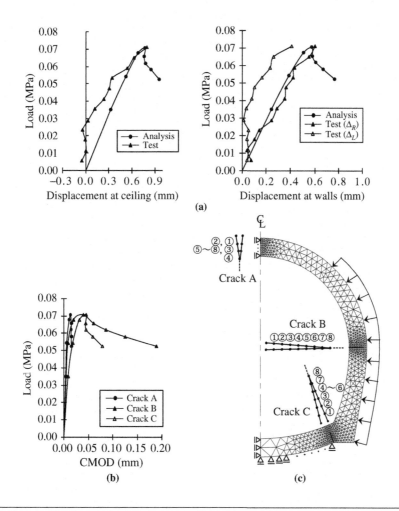

FIGURE 4.18 The results of a crack analysis by a half-FE-model: (a) load-displacement relation, (b) load-CMOD relation, and (c) crack propagation chart (after Shi et al., 2001; courtesy of ASCE).

4.6.3 Crack Analysis with a Full-FE-Model

Figure 4.19 presents a full-FE-model with five initial notches introduced into the tunnel specimen based on the exact crack locations observed during the experiment, as shown in Figure 4.16. The size of the initial notches is 25 mm, equal to one-tenth of the wall thickness. Since the ratio of the actual eccentric loads that occurred during the test was unknown, the uniformly distributed loads on the two sidewalls of the full-model are assumed to be equal, but the vertical supports under the right portion of the bottom plate are removed to simulate the unsymmetric boundary conditions of the test. Again, the crack paths are assumed to be normal to the lining surface. For the geometric details and the boundary conditions of the FE model, refer to Figure 4.19.

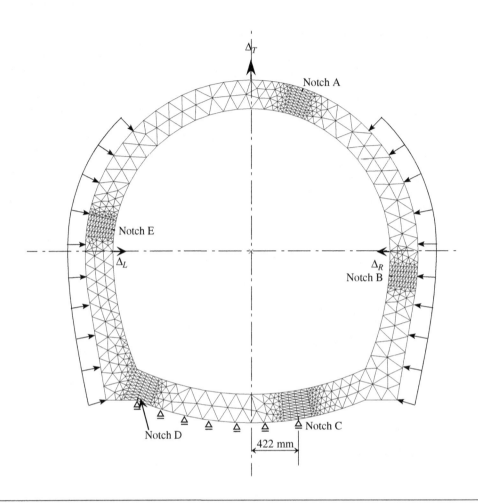

FIGURE 4.19 Full-FE-model of tunnel specimen.

Figure 4.20 presents the numerical results of load-displacement relations, load-CMOD relations, and crack propagation charts, which clearly demonstrate the effects of the unsymmetric boundary conditions (including the unsymmetric notch locations) on the structural response. The load-displacement relations at the ceiling and the sidewalls are compared with the test results. As seen, small differences are observed in the displacements of the sidewalls. The obtained displacements at the right wall are slightly larger than the test results, though the displacements at the left wall seem to be in reasonable agreement with the test measurements. From the crack propagation charts and the load-CMOD relations, crack B is shown to be the dominant crack, and its simultaneous propagation with crack E in the first four steps illustrates the vulnerability of the sidewalls against cracking. The growth of crack C in the bottom plate is slow but steady. Crack A and crack D are almost inactive, exhibiting exactly the same cracking behavior by moving only

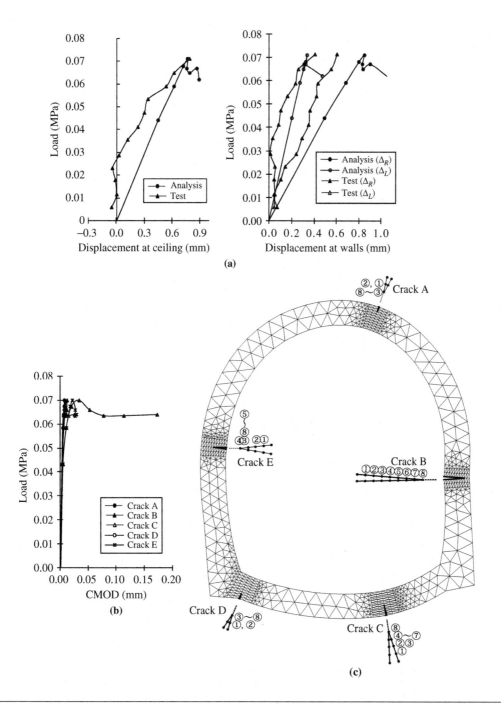

FIGURE 4.20 Results of crack analysis by a full-FE-model: (a) load-displacement relation, (b) load-CMOD relation, and (c) crack propagation chart.

one step forward and then becoming totally nonpropagating. Upon reaching the peak load at the fourth step, the active growth of crack E stops, and the crack becomes nonpropagating in the postpeak regions.

On the other hand, although crack C is less active compared to crack E in the prepeak region, it becomes a propagating crack again from the eighth step in the postpeak regions. The repropagation of crack C is closely related to the opening of crack B in the sidewall, which is accompanied by the large structural deformations of the tunnel specimen that in turn cause a large bending moment to form under the pressure loads at the right portion of the bottom plate to propagate crack C. In general, the numerically obtained cracking behaviors represent closely the crack propagation patterns of the test, except for crack D. The recorded large crack at notch D is not reproduced by the numerical analysis, probably due to the inaccuracy in the assumed boundary conditions for the actual situation.

Based on the preceding analyses of the two FE models, the obtained maximum CMOD at the sidewall is found to be approximately 0.2 mm before the crack penetrates the wall thickness and becomes a through crack. In real situations, it is not unusual to encounter aging waterway tunnels of similar scales that possess large cracks whose CMODs reach several millimeters. Obviously, the interactive support of the surrounding geological materials must be taken into account in crack analysis in order to allow such large cracks to develop in tunnel linings. A hybrid structural model is developed in Chapter 8 for this purpose, and combined with a loosening zone soil mechanics model, external earth pressures acting on tunnel linings can be numerically calculated based on the values of the actual CMODs.

4.7 CRACK ANALYSIS OF A SCALE-MODEL TEST OF A GRAVITY DAM BY CARPINTERI AND COLLEAGUES

In this section a fracture test on a scale model on a gravity dam by Carpinteri and colleagues will be presented, and the crack analysis of the test and of two additional cases will be discussed, focusing on the crack behavior of a dam with a single crack and with multiple cracks.

4.7.1 Background

Although monolithic structures such as dams are designed and shaped to withstand mainly compressive loads, most concrete dams present cracks in local tension zones that frequently become the cause of concern. In view of the grave consequences of a dam failure on human lives and the immense impact on society in general, accurate prediction of crack propagation and structural-stability assessment are of utmost necessity to ensure safe operation of the structures. In the following, the EFCM will be applied to study cracking behavior in a model dam to illustrate the effectiveness of the method in dealing with multiple cracks in dam structures.

Carpinteri et al. (1992) tested two 1:40 scale models of a concrete gravity dam subjected to equivalent hydraulic loads. The testing setup is shown in Figure 4.21. Each specimen contains a horizontal notch on the upstream face located at a quarter of the dam height, with a notch size of one-tenth of the dam thickness ($0.1W$) for the first specimen and one-fifth ($0.2W$) for the second one. Here, W represents the dam thickness at the location of a designated initial notch.

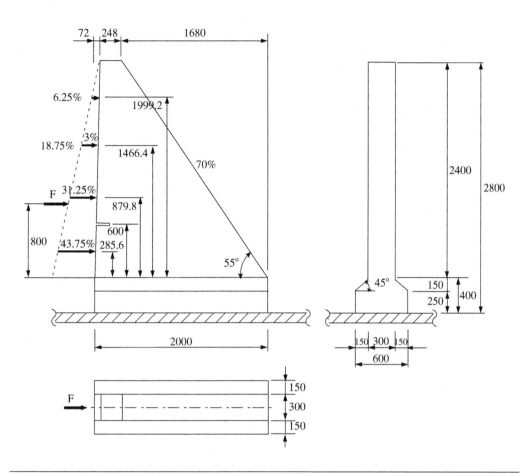

FIGURE 4.21 Model test of concrete gravity dam (dimensions in mm) by Carpinteri et al. (1992).

Following an unsuccessful experimental attempt to simulate the self-weight condition (Test 1), in which an unstable failure occurred along the base of the model, the repaired specimen was then fixed to the testing platform and loaded again until structural failure, without any self-weight simulation (Test 2). The fracture test of the second specimen was carried out under the same conditions (Test 3).

Figure 4.22 illustrates three numerical models: Model I, Model II, and Model III. In the following, Test 2 and Test 3 are analyzed first using Model I, and the predicted propagations of the single cracks are compared with the documented experimental and numerical analysis results of Carpinteri et al. Next, Model II and Model III are used to study crack propagations in concrete dams involving multiple cracks. Model II is of the same structure as Model I, but it contains two more initial notches in the upper part of the dam. Model III is a variation of the original structure of Model I and Model II, where the thickness of the dam is decreased in the upper part by removing the original top thickness and increased in the lower part by adding an upstream batter.

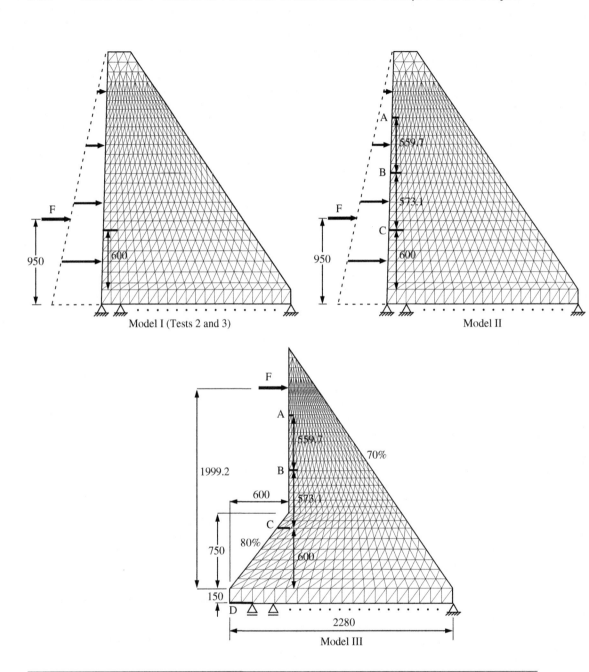

FIGURE 4.22 Three numerical models (dimensions in mm) of gravity dams with initial notches.

Table 4.3 Material Properties of a Dam Specimen

E (GPa)	v	γ (kg/m^3)	f_t (MPa)	G_F (N/mm)
35.70	0.10	2400	3.60	0.184

With four initial notches assumed and a single load applied near the top, this model is designed to increase the overall structural stability and thus allow more cracks to compete for steady growth, including the base crack, as well as to illustrate the dependence of cracking behavior on structural types. For the geometric details and loading conditions of each model, refer to Figure 4.22.

Although the uplift pressures are not considered in the following studies, they can be readily introduced to the surfaces of open cracks and treated like body forces. In solving the crack equations, the bilinear tension-softening relation in Figure 4.9 is again employed. In line with the common two-dimensional approach to dam structures, the numerical analyses are carried out under the plain strain conditions. The material properties of the test specimens and numerical models are listed in Table 4.3, where the unit weight of concrete, γ, is assumed to be 2400 kg/m^3.

4.7.2 **Model I: Single-Crack Propagation**

The experimental results of Test 2 and Test 3, as well as the numerical results of the present study and those of Carpinteri et al. are compared in Figure 4.23, which include the load-CMOD relations and crack propagation charts. As seen from the load-CMOD relations, both numerical analyses provide very close predictions of the ultimate resistance of the structure. For Test 2, the unexpectedly large CMODs of the test results in the postpeak regions seem to be attributed to the previous damage sustained by the notch of the first specimen during Test 1. For Test 3, a good agreement between the test and numerical analyses is observed.

Although the predicted crack paths of the present study match reasonably well with the actual crack paths during the steady crack growth, they diverge in the stage of unsteady crack propagation (marked distinctively by the kink in each obtained curve) shortly before structural failure. As shown in Figure 4.23, while the present study predicts that the cracks finally penetrate the model dams horizontally, the tests reveal that the actual paths curve toward the toes of the structures in the downstream side.

A careful study of the crack trajectories in the two tests, especially in Test 3, suggests that a transition of the loading conditions may have occurred before structural failure. Experimentally speaking, it may be extremely difficult to sustain the loading conditions as specified in Figure 4.21 during unsteady crack propagation. As the upper structure becomes unstable and starts to topple, the location of the resultant of the equivalent hydraulic loads may inevitably shift upward, thus releasing the compressive forces from the lower part of the structure and altering the stress fields to allow the cracks to curve downward. Without introducing such a change in the loading conditions as described, the numerically computed curvilinear crack path will inevitably bend horizontally because the new path leads to the minimum resistant moment of the remaining intact dam structure, thus leaving behind a kink in the crack path. Due to the limited information concerning their computational assumptions, further discussion of the numerical results of Carpinteri et al. on crack propagation is omitted.

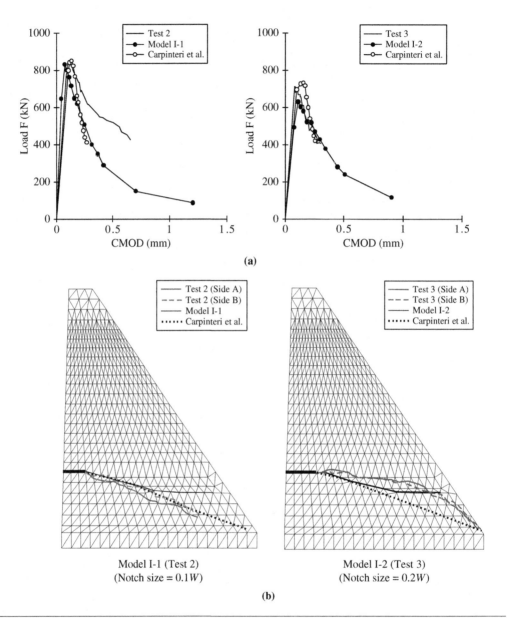

FIGURE 4.23 Numerical results of Model I: (a) load-CMOD relation and (b) crack propagation chart (after Shi et al., 2003; courtesy of ASCE).

It should be noted that due to the enormous self-weight of actual dams, the frictional forces on the crack surface might have a significant influence on crack propagation. By including shear, a mixed-mode fracture problem has to be solved, in which the inelastic material behavior of the fracture process zone is governed not only by the tension-softening law but also by a shear-transfer law as well. As will be discussed in Chapter 7 on mixed-mode fracture, by introducing a shear force to the crack surface, the computed crack path in Test 3 shifts upward and coincides with the actual crack path during the stable crack propagation.

4.7.3 **Model II: Multiple-Crack Propagation**

During the preliminary studies on Model II, it was found that by setting an initial notch (of any size) at the foundation in the upstream side, the base crack would inevitably become the dominant crack, allowing no cracking from the other notches in the dam at all. To allow other cracks to develop, the structure must be completely fixed at the bottom, assuming that no crack can propagate from there.

The numerical results of Model II are presented in Figure 4.24. As shown, three cases are considered: In Case 1 all the three initial notches are assigned the same size of 0.1W, and in Cases 2 and 3, notch B is enlarged to 0.3W and 0.4W, respectively, in order to activate crack B. From the numerical computations, crack C is found to be the only active crack in Case 1. This situation remains basically unchanged in Case 2, although competition from crack B for simultaneous propagation does occur briefly in the initial stage of crack growth. The transition of the cracking behavior finally takes place when an unrealistically large notch size of 0.4W is assumed for notch B in Case 3, and only then does crack B become the dominant crack. With its location further above crack B, the activation of crack A certainly requires an even larger initial size: 0.7W.

Compared with the cracking behaviors of the simple beams and tunnel specimens discussed previously, crack propagations in the present cases involve only limited crack interactions. In view of the structural characteristics of Model II, it is much easier for cracks to propagate from the lower sections of the structure than from the upper sections, because the dam thickness at the lower sections is not large enough to counterbalance the increase of the bending moment there. With a slender upper section and a sturdy lower part, more intense crack interactions can be expected in Model III.

4.7.4 **Model III: Multiple-Crack Propagation**

The numerical results of Model III are summarized in Figure 4.25. Unlike Model II, an initial notch is also introduced to the base of Model III, and in total four cases will be studied. While all four initial notches in Case 1 are assigned the same size of 0.1W, notch A of Case 2 and notch C of Case 3 are, respectively, enlarged to 0.2W, and notch D of Case 4 is given an even larger size of 0.4W for the base crack to extend.

With the four cases as described, four different kinds of cracking behavior are observed. In Case 1, crack B is the most critical crack, propagating horizontally while interacting with cracks A and C. With a larger notch size of 0.2W, crack A of Case 2 becomes the dominant crack, and its curvilinear trajectory turns sharply downward in the final stage of propagation. Although notch C and notch D are unaffected by the growth of crack A, interactions between crack A and crack B are evident: The sharp curving of crack A at the eighth step forces crack B to retreat (see Figure 4.25). In Case 3, a progressive crack grows out of notch C, curving upward while actively interacting with the two upper cracks, crack A and crack B. Despite their close geometric locations, interactions between crack C and notch D are absent. With the added upstream batter,

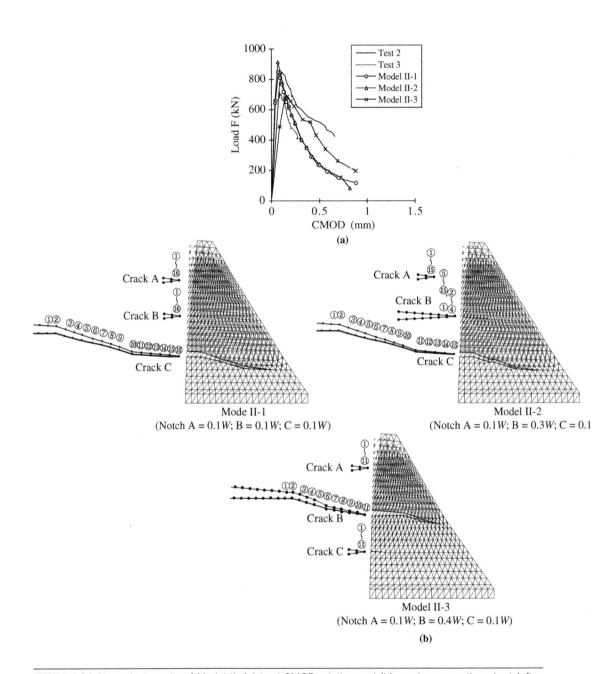

FIGURE 4.24 Numerical results of Model II: (a) load-CMOD relation and (b) crack propagation chart (after Shi et al., 2003; courtesy of ASCE).

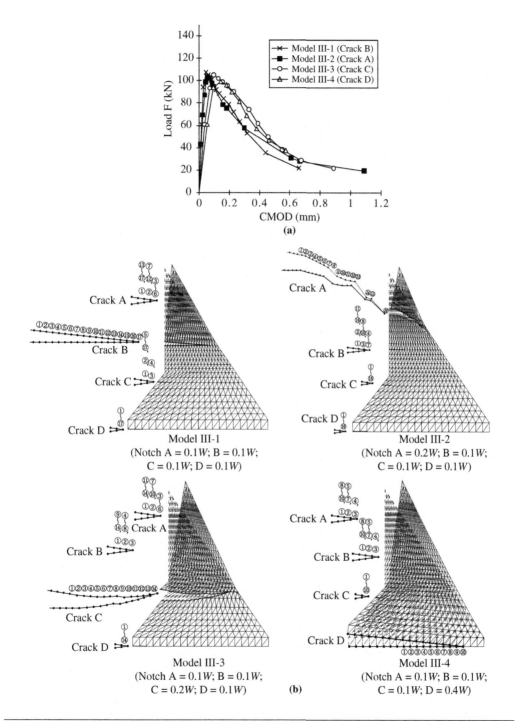

FIGURE 4.25 Numerical Results of Model III: (a) load-CMOD relation and (b) crack propagation chart (after Shi et al., 2003; courtesy of ASCE).

the structural stability is greatly enhanced at the foundation, and the base crack remains inactive until notch D is enlarged to 0.4*W*. Again, it is interesting to note that in Case 4, although crack D interacts quite actively with the two distant cracks, cracks A and B, the closest notch, notch C, seems to be unaffected. Obviously, these complicated interacting behaviors of the multiple cracks are closely related to the unique structural characteristics of Model III.

As clearly demonstrated through the numerical studies of Model II and Model III, with the present approach there is no need to limit the number of cracks in FE models of concrete dams. In general, the locations of initial defects or surface cracks in a dam should be determined through field investigations. However, they can also be introduced to any local tension zones in a dam based on a preliminary stress analysis or past experience. Unless these locations are indeed critical to crack propagation, no cracks will actually emerge from there.

As shown, the crack analysis using the EFCM provides an analytical tool that enables the ultimate response of concrete dams to be investigated and the potentially damaging cracks to be identified and their propagation paths predicted. Safety assessment of concrete dams using such a method serves as an additional check on the structural reliability of the dam designed by the traditional cantilever-beam theories. Furthermore, the information thus obtained can be used for devising effective remedial measures against further cracking of the existing cracks and thus reducing the potential risk of a dam failure.

REFERENCES

Abo, H., Tanaka, M., and Yoshida, N. (2000). "Development of a maintenance system for waterway tunnels." *Electric Power Civil Engineering*, 287, 42–46, JEPOC, Tokyo.

Carpinteri, A., Valente, S., Ferrara, G., and Imperato, L. (1992). "Experimental and numerical fracture modeling of a gravity dam." *Fracture Mechanics of Concrete Structures*, Z. P. Bazant, ed., pp. 351–360, Elsevier Applied Science.

Feng, L. M., Pekau, O. A., and Zhang, C. H. (1996). "Cracking analysis of arch dams by 3D boundary element method." *J. Struct. Eng.*, 122(6), 691–699.

Griffith, A. A. (1921). "The phenomena of rupture and flow in solids." *Phil. Trans. Roy. Soc. of London*, A 221, 163–197.

Griffith, A. A. (1924). "The theory of rupture." *Proc. 1st Int. Congress Appl. Mech.*, Biezeno and Burgers eds., pp. 55–63, Waltman.

Hillerborg, A., Modeer, M., and Petersson, P. E. (1976). "Analysis of crack formation and crack growth in concrete by means of fracture mechanics and finite elements." *Cement and Concrete Research*, 6(6), 773–782.

Ingraffea, A. R. (1990). "Case studies of simulation of fracture in concrete dams." *Eng. Fracture Mech.*, 35(1/2/3), 553–564.

Ohtsu, M. (1990). "Tension softening properties in numerical analysis." *Colloquium on Fracture Mechanics of Concrete Structures*, JCI Committee Report, 55–65.

Rokugo, K., Iwasa, M., Suzuki, T., and Koyanagi, W. (1989). "Testing methods to determine tensile strain softening curve and fracture energy of concrete." *Fracture Toughness and Fracture Energy-Test Method for Concrete and Rock*, H. Mihashi, H. Takahashi, and F. H. Wittmann eds., pp. 153–163, Balkema.

Shi, Z., Ohtsu, M., Suzuki, M., and Hibino, Y. (2001). "Numerical analysis of multiple cracks in concrete using the discrete approach." *J. Struct. Eng.*, 127(9), 1085–1091.

Shi, Z., Suzuki, M., and Nakano, M. (2003). "Numerical analysis of multiple discrete cracks in concrete dams using extended fictitious crack model." *J. Struct. Eng.*, 129(3), 324–336.

Uchida, Y., Rokugo, K., and Koyanagi, W. (1993). "Flexural tests of concrete beams." *Applications of Fracture Mechanics to Concrete Structures*, pp. 346–349, JCI Committee Report.

Zhang, C., and Karihaloo, B. L. (1992). "Stability of a crack in a large concrete dam." *Australian Civil Engineering Transactions*, IEAust., CE34(4), 369–375.

Crack Interaction and Localization

5.1 INTRODUCTION

Crack interaction is an important issue in the crack analysis of concrete. Because the local stress field and the crack driving force for a given flaw can be significantly affected by the presence of one or more neighboring cracks, clarifying the effect of crack interaction is the key to a clear understanding of various cracking behaviors, including crack localization. In linear elastic fracture mechanics (LEFM), it is known that depending on the relative orientation of the neighboring cracks, the crack interaction can either magnify or diminish the stress intensity factor.

To illustrate the point, consider two infinite plates under tension with two identical collinear cracks and two identical parallel cracks, respectively. Figure 5.1 presents the K_I solutions of the collinear cracks at the two crack tips, A and B, as the spacing, s, varies. As clearly shown, as the two cracks approach each other, the crack interaction magnifies the stress intensity factors at both crack tips, and the crack tip closest to the neighboring crack experiences a greater magnification in K_I. As $s \rightarrow 0$, the K_I solution at tip B increases asymptotically, while the solution at tip A approaches $\sqrt{2}$ because the two cracks coalesce into a single crack with twice the original length of each crack. The magnifying effect on K_I by the interaction of collinear cracks originates from the stress concentrations at the tips of each crack, which are tantamount to an increase in the applied load. The closer the crack tip to the neighboring crack, the greater the magnifying effect on K_I.

On the other hand, when the two cracks are parallel to each other, as illustrated in Figure 5.2, the stress intensity factor decreases relative to the single-crack case. The diminishing effect on K_I by the interaction of parallel cracks can be understood by considering the mutual shielding that results in a weaker stress field for each of the parallel cracks. These two cases highlight the complex nature and varying effects of crack interactions that must have a profound influence on various cracking behaviors, as will be revealed and discussed in the following sections.

Many studies have examined the interaction effects of multiple cracks in the fracturing process of concrete (Bazant and Wahab, 1980; Ingraffea et al., 1984; Barpi and Valente, 1998; Carpinteri and Monetto, 1999). In the discrete approach that allows the interaction of multiple cracks to be studied most straightforwardly, an explicit mathematical formulation of the crack interaction is possible (Shi et al., 2004). Such an approach enables crack interaction to be quantified and various cracking behaviors (such as why some cracks are active, while others are not and why crack localization begins early in some cases and is delayed in others) to be studied based on the nature

119

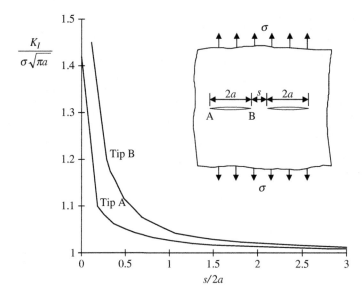

FIGURE 5.1 Interaction of two identical collinear cracks in an infinite plate (after Murakami, 1987).

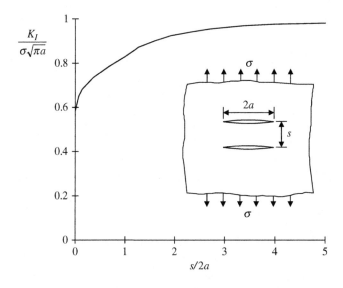

FIGURE 5.2 Interaction of two identical parallel cracks in an infinite plate (after Murakami, 1987).

and the intensity of the crack interactions involved. In order to derive the coefficient of interaction, a numerical formulation of three discrete cracks is carried out in the following section, based on the extended fictitious crack model (EFCM) developed in Chapter 4. Classifying the components of the tip force as the principal tip force and the secondary tip force, with the latter containing exclusively the tip force components caused by the neighboring cracks and thus reflecting the crack interaction, the coefficient of interaction is defined.

As revealed by the following studies, the nature and characteristics of crack interactions can vary greatly among various structural types that possess either one or multiple tension zones when deformed. Among a variety of studies on the fracture processes of concrete structures, concrete beams are undoubtedly one of the most frequently analyzed structural members. In fact, the first theoretical work on the fracture of concrete investigated the application of LEFM to concrete beams under three- and four-point bending (Kaplan, 1961). Although a concrete beam is structurally simple, a good understanding of its failure process, crack interaction, and localization may well reveal some of the fundamental mechanisms in the fracture of concrete structures. For this reason, two types of notched beams are studied, focusing on crack interaction and localization.

As the first group of the case study, the round-robin tests of unnotched plain concrete beams under four-point bending (Uchida et al., 1993) are selected. Introducing three small initial notches into the FE beam models with four different combinations of the notch sizes, the failure modes and the coefficients of interaction are obtained, and the mechanism of crack localization is analyzed. As the second group of the investigation, experimental and numerical studies on the fracture of plain concrete beams that contain both small and large initial notches are presented. To reveal the complex nature of crack interaction, a crack problem in a deformed tunnel lining structure, where several tension zones coexist, is chosen as the last group of the case study. Hence, one of the numerical models of the tunnel specimen discussed in Chapter 4 is employed to calculate the coefficients of interaction, with which the cracking behavior in the tunnel specimen is examined from a different perspective.

5.2 COEFFICIENT OF INTERACTION

In this section the crack equations for three arbitrary cracks are established first, which are used to analyze the source of crack iteration and introduce the coefficient of interaction.

5.2.1 Crack Equations and the Source of Crack Interaction

Since the coefficient of interaction will be derived from crack equations, the numerical procedure for solving three discrete cracks is outlined following. Figure 5.3 illustrates three cracks of the mode-I type: crack A, crack B, and crack C. Notice that a crack propagates in the direction normal to the tensile force at the tip of the fictitious crack. For clarity, the forces and crack-opening displacements (CODs) of the inactive cracks are marked by asterisks. If crack A is assumed to be the only propagating crack, the tensile force at its tip must reach the nodal force limit Q_{la}, as shown in Figure 5.3e,

$$Q_{la} = CR_a \cdot P_a + \sum_{i=1}^{N1} CI_{aa}^i F_a^i + \sum_{j=1}^{N2} CI_{ab}^j F_b^{*j} + \sum_{l=1}^{N3} CI_{ac}^l F_c^{*l} \tag{5.1}$$

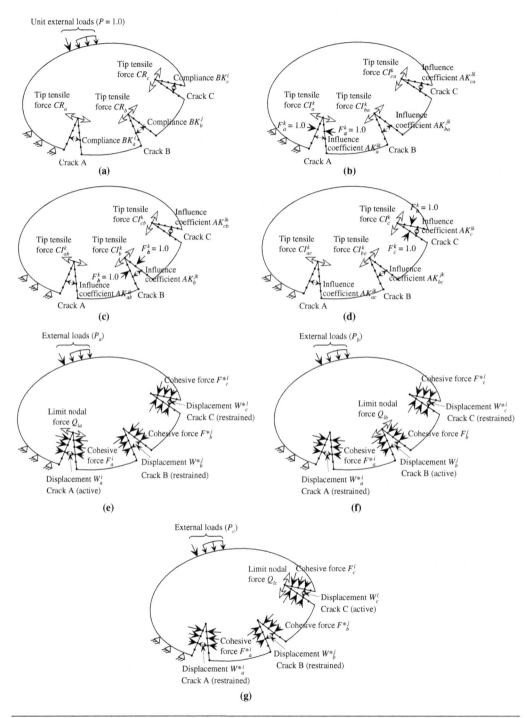

FIGURE 5.3 Crack-tip-controlled modeling of three discrete cracks: Forces and displacements at the cracks due to (a) unit external loads; (b) a pair of unit cohesive forces at crack A; (c) a pair of unit cohesive forces at crack B; and (d) a pair of unit cohesive forces at crack C. Load conditions for the growth of (e) crack A, (f) crack B, and (g) crack C.

where N1, N2, and N3 are the number of nodes inside the three fictitious cracks, respectively. Here, CR_a, CI^i_{aa}, CI^j_{ab}, and CI^l_{ac} represent the tensile forces at the tip of crack A due to a unit external load, a pair of unit cohesive forces at the ith node of crack A, the jth node of crack B, and the lth node of crack C, respectively. These coefficients are determined by FE calculations of the models in Figure 5.3a–d. Notice that P_a is the load required to propagate crack A, while crack B and crack C remain inactive.

As seen, the first term on the right-hand side of Eq. (5.1) represents the tip tensile force caused by the external loads, the second term represents the tip compressive force due to the cohesive forces at crack A itself, and the remaining terms stand for the tip force components caused by the cohesive forces of the neighboring cracks. The tip force component due to a neighboring crack can be either tensile or compressive, depending on the relative location of that crack to crack A and the deformation characteristics of the structural type involved.

The CODs along the three fictitious cracks are given by

$$W^i_a = BK^i_a \cdot P_a + \sum_{k=1}^{N1} AK^{ik}_{aa} F^k_a + \sum_{j=1}^{N2} AK^{ij}_{ab} F^{*j}_b + \sum_{l=1}^{N3} AK^{il}_{ac} F^{*l}_c \tag{5.2}$$

$$W^{*j}_b = BK^j_b \cdot P_a + \sum_{i=1}^{N1} AK^{ji}_{ba} F^i_a + \sum_{k=1}^{N2} AK^{jk}_{bb} F^{*k}_b + \sum_{l=1}^{N3} AK^{jl}_{bc} F^{*l}_c \tag{5.3}$$

$$W^{*l}_c = BK^l_c \cdot P_a + \sum_{i=1}^{N1} AK^{li}_{ca} F^i_a + \sum_{j-1}^{N2} AK^{lj}_{cb} F^{*j}_b + \sum_{k=1}^{N3} AK^{lk}_{cc} F^{*k}_c \tag{5.4}$$

where $i = 1, \ldots, N1$; $j = 1, \ldots, N2$; and $l = 1, \ldots, N3$. Here, BK^i_a at crack A, BK^j_b at crack B, and BK^l_c at crack C are the compliances at nodes i, j, and l, respectively, due to the external load. The influence coefficients AK^{ik}_{aa}, AK^{ij}_{ab}, and AK^{il}_{ac} are the displacements at the ith node of crack A due to a pair of unit cohesive forces at the kth node of crack A, the jth node of crack B, and the lth node of crack C, respectively.

Similarly, the influence coefficients AK^{ji}_{ba}, AK^{jk}_{bb}, and AK^{jl}_{bc} represent the displacements at the jth node of crack B, and AK^{li}_{ca}, AK^{lj}_{cb}, and AK^{lk}_{cc} are the displacements at the lth node of crack C, respectively, due to a pair of unit cohesive forces at the corresponding locations. FE models to compute these coefficients are given in Figures 5.3a–d. Based on Eqs. (5.2)–(5.4), the CODs, which represent the change of geometric shapes due to crack propagation, are functions of the external loads, the cohesive forces at the given crack, and the cohesive forces at the surrounding cracks.

Imposing the tension-softening law of concrete along each fictitious crack leads to

$$F^i_a = f(W^i_a) \tag{5.5}$$

$$F^{*j}_b = f(W^{*j}_b) \tag{5.6}$$

$$F^{*l}_c = f(W^{*l}_c) \tag{5.7}$$

Eqs. (5.1) through (5.7) form the crack equations for the case in which crack A is the sole propagating crack. With the number of equations (2N1 + 2N2 + 2N3 + 1) matching the number of unknowns (2N1 + 2N2 + 2N3 + 1), the problem can be solved uniquely.

The crack equations for propagating crack B are obtained as

$$Q_{lb} = CR_b \cdot P_b + \sum_{i=1}^{N1} CI_{ba}^i F_a^{*i} + \sum_{j=1}^{N2} CI_{bb}^j F_b^j + \sum_{l=1}^{N3} CI_{bc}^l F_c^{*l} \tag{5.8}$$

$$W_a^{*i} = BK_a^i \cdot P_b + \sum_{k=1}^{N1} AK_{aa}^{ik} F_a^{*k} + \sum_{j=1}^{N2} AK_{ab}^{ij} F_b^j + \sum_{l=1}^{N3} AK_{ac}^{il} F_c^{*l} \tag{5.9}$$

$$W_b^j = BK_b^j \cdot P_b + \sum_{i=1}^{N1} AK_{ba}^{ji} F_a^{*i} + \sum_{k=1}^{N2} AK_{bb}^{jk} F_b^k + \sum_{l=1}^{N3} AK_{bc}^{jl} F_c^{*l} \tag{5.10}$$

$$W_c^{*l} = BK_c^l \cdot P_b + \sum_{i=1}^{N1} AK_{ca}^{li} F_a^{*i} + \sum_{j=1}^{N2} AK_{cb}^{lj} F_b^j + \sum_{k=1}^{N3} AK_{cc}^{lk} F_c^{*k} \tag{5.11}$$

$$F_a^{*i} = f(W_a^{*i}) \tag{5.12}$$

$$F_b^j = f(W_b^j) \tag{5.13}$$

$$F_c^{*l} = f(W_c^{*l}) \tag{5.14}$$

The crack equations for propagating crack C are given by

$$Q_{lc} = CR_c \cdot P_c + \sum_{i=1}^{N1} CI_{ca}^i F_a^{*i} + \sum_{j=1}^{N2} CI_{cb}^j F_b^{*j} + \sum_{l=1}^{N3} CI_{cc}^l F_c^l \tag{5.15}$$

$$W_a^{*i} = BK_a^i \cdot P_c + \sum_{k=1}^{N1} AK_{aa}^{ik} F_a^{*k} + \sum_{j=1}^{N2} AK_{ab}^{ij} F_b^{*j} + \sum_{l=1}^{N3} AK_{ac}^{il} F_c^l \tag{5.16}$$

$$W_b^{*j} = BK_b^j \cdot P_c + \sum_{i=1}^{N1} AK_{ba}^{ji} F_a^{*i} + \sum_{k=1}^{N2} AK_{bb}^{jk} F_b^{*k} + \sum_{l=1}^{N3} AK_{bc}^{jl} F_c^l \tag{5.17}$$

$$W_c^l = BK_c^l \cdot P_c + \sum_{i=1}^{N1} AK_{ca}^{li} F_a^{*i} + \sum_{j=1}^{N2} AK_{cb}^{lj} F_b^{*j} + \sum_{k=1}^{N3} AK_{cc}^{lk} F_c^k \tag{5.18}$$

$$F_a^{*i} = f(W_a^{*i}) \tag{5.19}$$

$$F_b^{*j} = f(W_b^{*j}) \tag{5.20}$$

$$F_c^l = f(W_c^l) \tag{5.21}$$

By solving the three sets of crack equations, the true cracking mode and crack paths for the next load increment are determined based on the minimum load criterion, and the stress and displacement fields are calculated accordingly. This process is repeated until the divergence of numerical solutions occurs at structural failure. Detailed descriptions on correcting invalid solutions for relevant cracking modes can be found in Chapter 4.

The crack equations govern the change of the geometric shapes and the alteration of the stress and displacement fields due to crack propagation and are directly related to three types of forces:

the external loads, the cohesive forces at the given crack, and the cohesive forces at the neighboring cracks. Obviously, the so-called crack interaction refers specifically to any effect that has a definite influence on the cracking behavior and is caused by the cohesive forces of the neighboring cracks. Thus, it seems relevant to derive a coefficient of interaction from the crack equations. This coefficient may provide a quantitative measure of the crack interactions for a specific cracking behavior at a given step of crack propagation. The coefficient of interaction will be derived in the following section.

5.2.2 Coefficient of Interaction and Principal Tip Force Coefficient

Because the tip force is caused by the external loads and the cohesive forces of all the cracks involved, the interference effect between cracks may be measured by calculating the ratio of the tip force components due to the neighboring cracks to the nodal force limit. Accordingly, the nodal force components at the tip of an active crack (see Eqs. (5.1), (5.8), and (5.15)) are divided into two parts: the principal tip force (PTF) Q_a^I, Q_b^I, and Q_c^I given by

$$Q_a^I = CR_a \cdot P_a + \sum_{i=1}^{N1} CI_{aa}^i F_a^i \qquad (5.22)$$

$$Q_b^I = CR_b \cdot P_b + \sum_{j=1}^{N2} CI_{bb}^j F_b^j \qquad (5.23)$$

$$Q_c^I = CR_c \cdot P_c + \sum_{l=1}^{N3} CI_{cc}^l F_c^l \qquad (5.24)$$

and the secondary tip force (STF) Q_a^{II}, Q_b^{II}, and Q_c^{II}, given by

$$Q_a^{II} = \sum_{j=1}^{N2} CI_{ab}^j F_b^{*j} + \sum_{l=1}^{N3} CI_{ac}^l F_c^{*l} \qquad (5.25)$$

$$Q_b^{II} = \sum_{i=1}^{N1} CI_{ba}^i F_a^{*i} + \sum_{l=1}^{N3} CI_{bc}^l F_c^{*l} \qquad (5.26)$$

$$Q_c^{II} = \sum_{i=1}^{N1} CI_{ca}^i F_a^{*i} + \sum_{j=1}^{N2} CI_{cb}^j F_b^{*j} \qquad (5.27)$$

where

$$Q_{ta} = Q_a^I + Q_a^{II} \qquad (5.28)$$

$$Q_{tb} = Q_b^I + Q_b^{II} \qquad (5.29)$$

$$Q_{tc} = Q_c^I + Q_c^{II} \qquad (5.30)$$

As seen, the tip force components of the PTF result from the external loads and the cohesive forces at the active crack itself, whose resultant is invariably a tensile force. On the other hand, the STF contains exclusively the tip force components caused by the cohesive forces at the surrounding cracks and therefore represents the crack interactions. Depending on the specific configuration of the problem and the relative locations of the neighboring cracks to the active crack, the resultant of the STF components can either be a tensile force or a compressive force. Hence, the interactions of the neighboring cracks may facilitate or hinder the propagation of the active crack, depending on whether the STF is tensile or compressive (notice the similarities with the magnifying and diminishing effects of the crack interaction on the stress intensity factors of the collinear and parallel cracks in LEFM). To introduce the coefficient of interaction and the PTF coefficient, the PTF and the STF are now divided by the critical tip force, and the resulting non-dimensionalized coefficients μ_a^I, μ_b^I, and μ_c^I for the PTF, and μ_a^{II}, μ_b^{II}, and μ_c^{II} for the STF, are presented as

$$\mu_a^I = \frac{Q_a^I}{Q_{Ia}} \tag{5.31}$$

$$\mu_b^I = \frac{Q_b^I}{Q_{Ib}} \tag{5.32}$$

$$\mu_c^I = \frac{Q_c^I}{Q_{Ic}} \tag{5.33}$$

$$\mu_a^{II} = \frac{Q_a^{II}}{Q_{Ia}} \tag{5.34}$$

$$\mu_b^{II} = \frac{Q_b^{II}}{Q_{Ib}} \tag{5.35}$$

$$\mu_c^{II} = \frac{Q_c^{II}}{Q_{Ic}} \tag{5.36}$$

where

$$\mu_a^I + \mu_a^{II} = 1 \tag{5.37}$$

$$\mu_b^I + \mu_b^{II} = 1 \tag{5.38}$$

$$\mu_c^I + \mu_c^{II} = 1 \tag{5.39}$$

Here, μ_a^{II}, μ_b^{II}, and μ_c^{II} are termed the coefficients of interaction, and μ_a^I, μ_b^I, and μ_c^I are called the PTF coefficients. Based on the previous analysis of the PTF and the STF, it is known that while μ_a^I, μ_b^I, and μ_c^I are always positive, μ_a^{II}, μ_b^{II}, and μ_c^{II} can be either positive or negative. Although the coefficient of interaction is derived from the tip force components alone, it reflects all the physical characteristics that a neighboring crack possesses through the cohesive forces of the crack. The coefficients of interaction will be employed in the following numerical studies on the fracture of beams and tunnel linings. As will be demonstrated, with such a coefficient to monitor and quantify crack interactions, cracking behavior can be analyzed and understood better in view of the delicate interaction effects involved. Ultimately, this will help clarify the various mechanisms of crack localization.

5.3 CRACK INTERACTIONS IN NOTCHED CONCRETE BEAMS UNDER FOUR-POINT BENDING

This section presents detailed analysis on crack interactions in notched concrete beams, assuming beams with only small notches and beams with both small and large notches.

5.3.1 Beams with Small Notches

Crack interactions in beams with only small notches are discussed in the following sections.

Numerical Models and Material Properties

The four-point bending test of the unnotched plain concrete beam, which was analyzed in Chapter 4 as multiple-crack problems by introducing several initial notches into FE beam models, is illustrated in Figure 5.4 along with four FE models. As shown, the four cases of the simple beam, Cases 1 through 4, contain three small initial notches. While the two side-notches A and C are

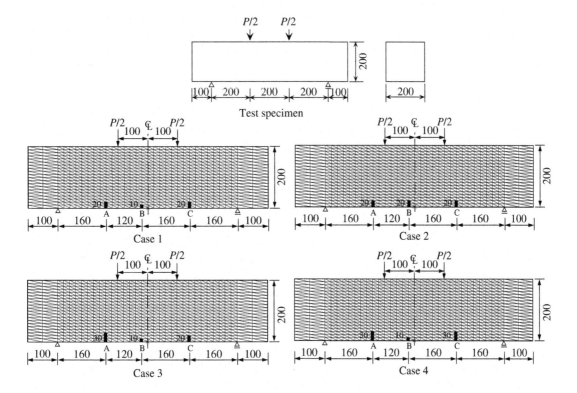

FIGURE 5.4 Fracture test (dimensions in mm) of an unnotched plain concrete beam and four numerical models with small initial notches.

symmetrically located with respect to the center of the beam, notch B is set slightly away from the center to introduce imperfection to the problem. Their respective locations are kept the same in all four cases. By varying the sizes of these initial notches, various cracking behaviors are obtained.

For comparison, the four cases are divided into two groups: the first group of Case 1 and Case 2, and the second group of Case 3 and Case 4. All geometric details are illustrated in Figure 5.4. As seen, the crack increment is set to 10 mm, which is one-twentieth of the beam depth. According to the preliminary study on mesh-size sensitivity, as the mesh size was reduced from 10 mm to 5 mm, the maximum load decreased slightly and the range of decrement was within 3 percent. To solve the crack equations, the bilinear tension-softening relation, as shown in Figure 4.9, is employed. The material properties of the test specimen are given in Table 4.1.

Results and Discussion

Figures 5.5 and 5.6 summarize the numerical results of the load-displacement relations, the load-CMOD relations, the stress contours at the final computational steps, and the crack propagation charts for the two groups of cases, respectively. To facilitate the discussion on crack interactions, Figures 5.7 to 5.10 exhibit, for each of the four cases respectively, the required loads, the coefficients of interaction, and the components of the PTF at each computational step for each assumed active crack. Notice that at each step of the computation, the true active crack is determined based on the minimum load criterion, and stress analysis is carried out accordingly.

In the first two cases, notches A and C are given the same size of 20 mm, and notch B is enlarged from 10 mm in Case 1 to 20 mm in Case 2. As shown in Figure 5.5, the maximum loads predicted in the two cases are slightly lower than the experimental results, which is understandable because the fracture tests were performed on unnotched specimens. Obviously, the maximum loads of the numerical models depend on the sizes of the initial notches. The peak load in Case 2 is decreased by approximately 10 percent compared with Case 1, due to the 10 mm increase in the size of notch B.

While the stress contours clearly show the stress concentrations at the crack tips, the propagation charts illustrate the detailed process of crack propagation. In Case 1, all three cracks compete for simultaneous propagation in the early stage of load increase. As the load approaches its maximum, cracks A and C stop growing at the fifth and sixth steps, respectively, leaving crack B as the only active crack. The peak load is reached at the sixth step, and eventually the two side cracks close as crack B penetrates deeper into the beam in the postpeak regions. In Case 2, due to a larger initial size, crack B encounters less competition for simultaneous propagation than the two side cracks. An early sign of crack localization is observed at crack B, when the propagations of cracks A and C are temporarily interrupted at initial steps. Again, the maximum load is obtained at the sixth step. Increasing the initial size of notch B further, the two side cracks will eventually lose any chance to propagate, and the crack localization will take place from the very beginning of the loading process.

Next, the fracturing processes just described are analyzed in terms of the crack interactions. As shown in Figure 5.7 of Case 1, the required loads for propagating each single crack are almost identical in the prepeak region. Under these circumstances, the subsequent stress analysis often leads to situations in which the tensile forces at the tips of the restrained cracks exceed the tensile strength of concrete. This explains the simultaneous propagation of all three cracks in the prepeak region in Case 1, as a result of modifying the assumed single-active-crack modes. The obtained

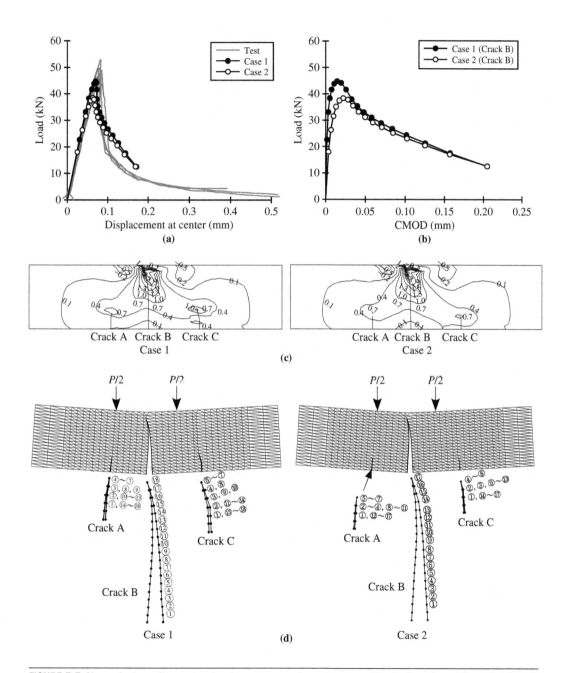

FIGURE 5.5 Numerical results on structural response and crack propagation in Case 1 and Case 2: (a) load-displacement relation, (b) load-CMOD relation, (c) maximum principal stress contours (MPa), and (d) crack propagation charts.

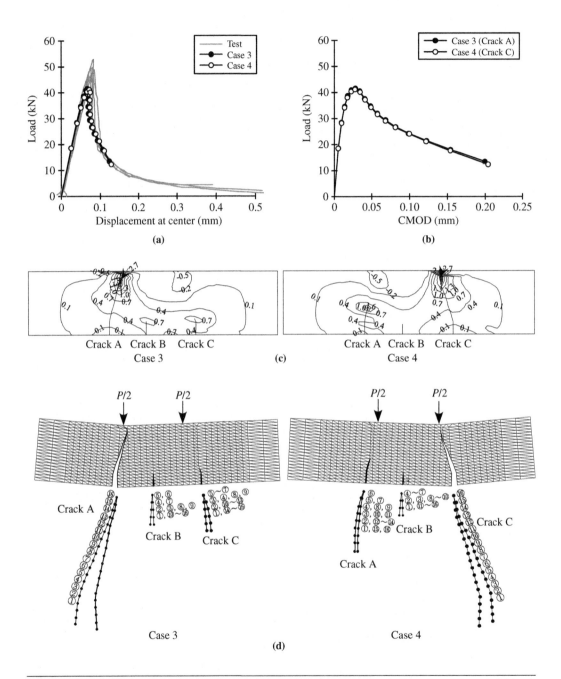

FIGURE 5.6 Numerical results on structural response and crack propagation in Case 3 and Case 4: (a) load-displacement relation, (b) load-CMOD relation, (c) maximum principal stress contours (MPa), and (d) crack propagation charts.

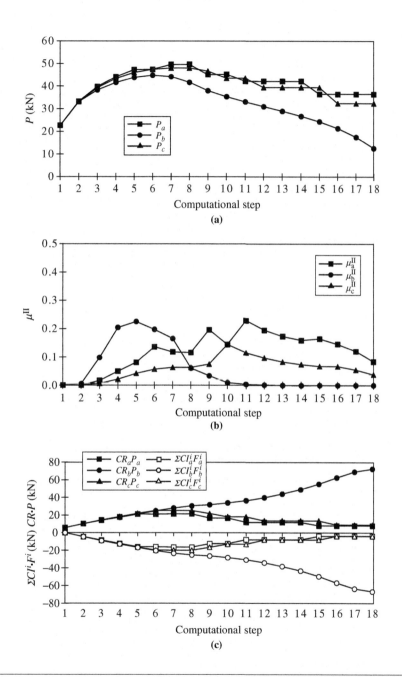

FIGURE 5.7 (a) External loads, (b) coefficients of interaction, and (c) principal tip force components for each assumed active crack (Case 1) (after Shi et al., 2004; courtesy of JCI).

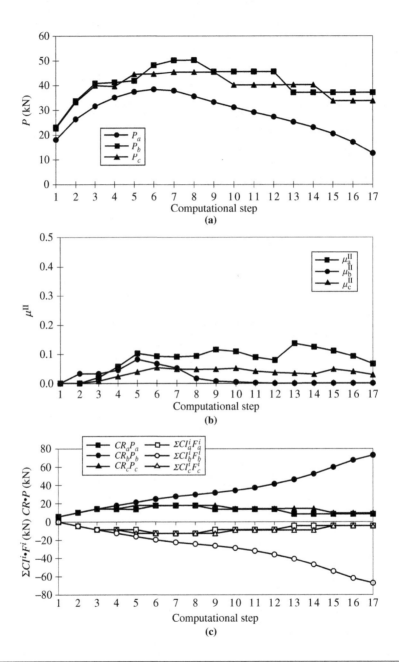

FIGURE 5.8 (a) External loads, (b) coefficients of interaction, and (c) principal tip force components for each assumed active crack (Case 2) (after Shi et al., 2004; courtesy of JCI).

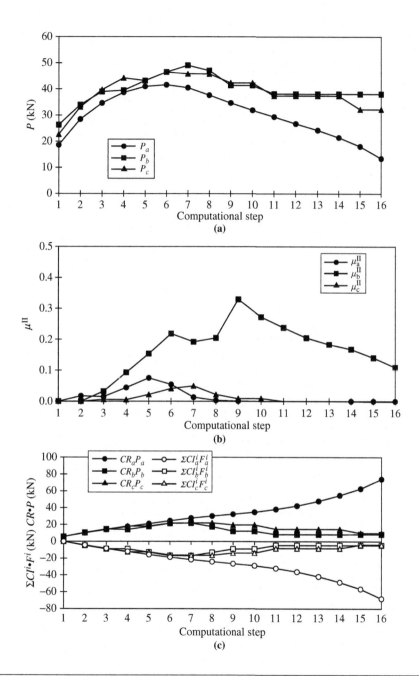

FIGURE 5.9 (a) External loads, (b) coefficients of interaction, and (c) principal tip force components for each assumed active crack (Case 3) (after Shi et al., 2004; courtesy of JCI).

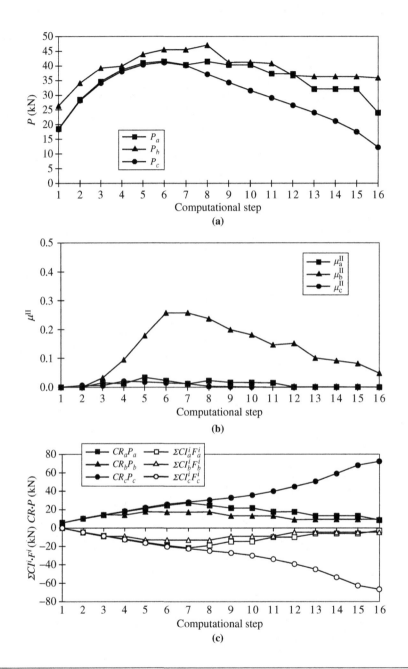

FIGURE 5.10 (a) External loads, (b) coefficients of interaction, and (c) principal tip force components for each assumed active crack (Case 4) (after Shi et al., 2004; courtesy of JCI).

coefficients of interaction at crack B, $\mu_b{}^{II}$, reveal an active role of crack interactions by cracks A and C in the pre- and postpeak regions. Since these coefficients are positive (except for an extremely small negative value recorded at the second computational step), the interactions contribute to increasing the tip tensile force and effectively facilitating the growth of crack B. The coefficient $\mu_b{}^{II}$ attains its maximum with the last simultaneous propagation of cracks B and C at the fifth step (refer to the propagation chart of Case 1 in Figure 5.5) and decreases afterward.

As crack B penetrates beyond half of the beam depth, $\mu_b{}^{II}$ becomes zero, and thus crack interactions are no longer involved in the subsequent growth of crack B. With extensive cracking concentrated at notch B, the required load for propagating crack B is greatly reduced. In contrast, $\mu_a{}^{II}$ and $\mu_c{}^{II}$ exhibit persistent crack interactions in the postpeak regions, and the required loads for crack propagation in the nonlocalized cracking modes, P_a and P_c, are much greater than that in P_b, obviously because a greater amount of the fracture energy is required in these situations than to propagate a single localized crack, crack B. Due to the crack localization, the stress concentration at the tip of crack B gradually intensifies, which is evident from the two components of the PTF that are in sharp contrast to those of the other modes.

Compared with Figure 5.7, the reductions in both the required loads P_b and the coefficients of interaction $\mu_a{}^{II}$, $\mu_b{}^{II}$, and $\mu_c{}^{II}$ in Figure 5.8 of Case 2 are apparent. These smaller coefficients of interaction reflect a more independent characteristic of the cracking behavior of the assumed active crack in Case 2 than in Case 1. As a result of assuming a larger notch size, the crack localization at crack B begins early, which is evidenced by the quickly diminishing values of $\mu_b{}^{II}$ in the postpeak regions as compared to those of Case 1. For details of the numerical results, refer to Figure 5.8.

We now address Case 3 and Case 4, focusing on the cracking behaviors. In these two cases, notches A and B are assigned the same sizes of 30 mm and 10 mm, respectively. Notch C is enlarged from 20 mm in Case 3 to 30 mm in Case 4. As shown in Figure 5.6, with a 10 mm difference in the size of notch C, the patterns of crack propagation and the stress distributions completely reverse, even though the maximum loads are virtually unchanged in the two cases. In Case 3, crack A is most active, suppressing the growth of both cracks B and C in the prepeak region. In Case 4, however, cracks A and C propagate simultaneously up to the peak load at the sixth step, reaching almost one-half of the beam depth with no sign of crack localization. The localization of cracking occurs at the seventh step when crack C finally emerges as the dominant crack, forcing the others to eventually close in the postpeak regions. The converse of the cracking behaviors in these two cases raises interesting questions on the role of crack interaction in crack localization.

The contrasting cracking behaviors just described are well mirrored in each of the three sets of relations in Figures 5.9 and 5.10. For example, the progressiveness of crack A in Case 3 is explained by the fact that P_a is the minimum required load from the very beginning of the loading process, as shown in Figure 5.9. Also, the intense competition for simultaneous propagation between cracks A and C in Case 4 is clearly attributable to the identical loads P_a and P_c in the first seven steps, as shown in Figure 5.10. In both cases the coefficients of interaction $\mu_a{}^{II}$ and $\mu_c{}^{II}$ are rather small. Nevertheless, the crack interaction in each case has played an important role in the process of crack localization. In Case 3 the propagation of crack A clearly benefits from the crack interactions in the prepeak region ($\mu_a{}^{II} > \mu_c{}^{II}$) that in turn accelerate the crack localization in the postpeak regions, where $\mu_a{}^{II}$ diminishes faster than $\mu_c{}^{II}$ does. In Case 4, the simultaneous propagation of crack A with crack C is disrupted because the persistent interactions as represented by $\mu_a{}^{II}$ in the postpeak regions have prevented crack A from becoming a major localizing crack. The lack of activity in crack B in Cases 3 and 4 is well reflected in the large values of $\mu_b{}^{II}$, which are mainly due to the small size of notch B and the geometric proximity of notch B to the two side notches. Again, the stress concentrations at crack A of Case 3 and crack C of Case 4 are observed in the two components of the PTF in Figures 5.9 and 5.10.

5.3.2 **Beams with Both Small and Large Notches**

Crack interactions in beams with both small and large notches are discussed in the following.

Aim of the Test

Since the fracture of a plain concrete beam under bending is quasi-brittle, it is difficult to detect crack propagations in the beam during a fracture test by visual inspection prior to the beam failure. Throughout the discussions of Cases 1 to 4, it has been demonstrated that the effect of crack interaction can be clearly identified in various cracking behaviors, including crack localization. To provide experimental evidence on the crack activities at multiple initial notches, four-point bending tests on notched simple beams were carried out, and the crack-mouth-opening displacement (CMOD) at each of the notches was carefully recorded. The test arrangement and the notch arrangement are illustrated in Figure 5.11, along with two FE models. As shown, three small notches of 10 mm were cut below the two loading points (notches A and C) and at the midspan (notch B), and a large notch of 50 mm (notch D; one-fourth of the beam depth) was further introduced into the beam in the middle of the distance between notch C and the nearest support. The width of the cuts was 4 mm. The concrete composition is shown in Table 5.1, and the material properties are listed in Table 5.2. Prior to the experiments, numerical analyses were performed to predict cracking behavior, using the FE beam models of Cases 5 and 6 in Figure 5.11.

Numerical studies showed that by assuming notch D to be 40 mm in Case 5, the dominant crack developed from the center. Upon increasing the size of notch D to 50 mm in Case 6, however, the most active crack originated from notch D. These two types of failure were both observed during the fracture tests—that is, some of the specimens broke at the midspan, while the rest failed at notch D. The variations of the test results are understandable in view of the inherent material inhomogeneities of concrete, the unavoidable small fluctuations in notch processing (as regarding the size and the position of a notch), and all the slight inaccuracies possibly encountered during the loading process. It is worth emphasizing that all of the notches' active crack openings were monitored.

Results and Discussion

The experimental results are divided into two groups based on the failure modes and are compared with the numerical results of Case 5 in Figure 5.12 and Case 6 in Figure 5.13, respectively. Based on the maximum values of the recorded CMODs at the peak load, the most active crack in Figure 5.12 was crack B (0.041 mm), followed by crack D (0.018 mm), crack A (0.010 mm), and crack C (0.007 mm). Similarly, the most active crack in Figure 5.13 was crack D (0.053 mm), followed by crack B (0.023 mm), crack A (0.017 mm), and crack C (0.012 mm).

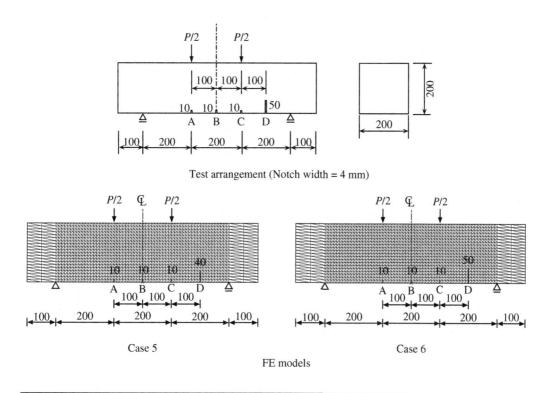

FIGURE 5.11 Four-point bending test (dimensions in mm) of plain concrete beams with small and large initial notches and two FE models.

Table 5.1 Concrete Composition of a Simple Beam with a Large Notch

Maximum Size of Coarse Aggregate (mm)	Slump (cm)	Air (%)	W/C Ratio (%)	Sand-Coarse Aggregate Ratio (%)	Unit Weight (kg/m³)				Unit Weight (g/m³)
					Water W	Cement C	Fine Aggregate S	Coarse Aggregate G	Admixture AE
20	11.5	4.5	55.0	46.0	165	300	834	991	9.00

Table 5.2 Material Properties of a Simple Beam with a Large Notch

E (GPa)	v	f_c (MPa)	f_t (MPa)	G_F (N/mm)
29.25	0.18	42.25	3.66	0.14

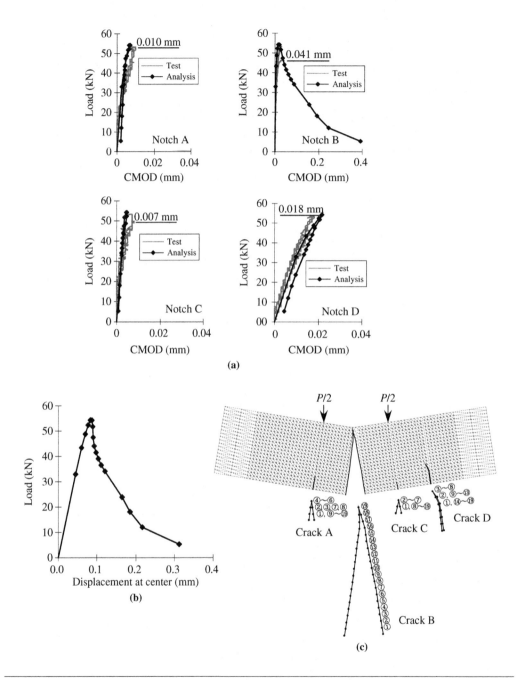

FIGURE 5.12 Experimental and numerical results on (a) load-CMOD relation, (b) load-displacement relation, and (c) crack propagation chart (Case 5).

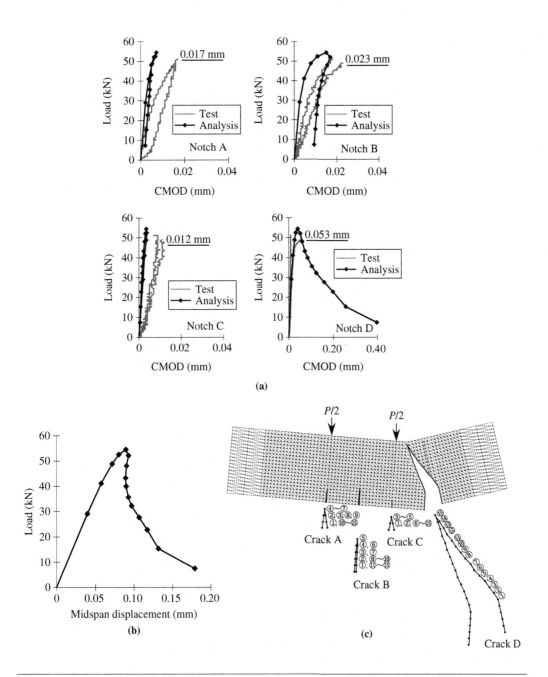

FIGURE 5.13 Experimental and numerical results on (a) load-CMOD relation, (b) load-displacement relation, and (c) crack propagation chart (Case 6).

Judging from the numerically obtained load-CMOD relations, it can be concluded that the agreement between the tests and the analyses is very good in Figure 5.12 and fairly good in Figure 5.13, which justifies the calculation of the coefficients of interaction and the following discussions on the cracking behavior and the crack localization with the help of these coefficients. It should also be noticed from the crack propagation charts in Figures 5.12 and 5.13 that the general cracking behaviors obtained numerically are in perfect agreement with the previous analyses of the recorded load-CMOD relations.

As shown in the propagation charts of Figure 5.12 and the load-step relations of Figure 5.14, with the size of notch D equal to 40 mm in Case 5, crack B is found to be the most active crack based on the minimum load criterion. After the initial simultaneous propagations of all four cracks at the second step (as the result of modifying the crack propagation patterns due to the stress concentrations at cracks A, C, and D), crack D becomes the only competing crack with crack B at the third step. This may seem a little puzzling at first because the required load to activate crack D alone is even greater than the load required to propagate crack A, as shown in Figure 5.14. When taking the large size of notch D into consideration, however, clearly the stress concentration at crack D due to the propagation of crack B must be greater than the stress concentration at crack A, thus resulting in the simultaneous propagation of crack D (not crack A) at the third step. This exemplifies the size of a large notch or crack as an influencing factor on the cracking behavior.

At the next step, the obtained P_a is only slightly larger than P_b. After modifying the crack propagation patterns based on the results of stress analysis, crack A is found to extend with crack B, while cracks C and D remain inactive. Compared with the small coefficients of μ_d^{II} in Figure 5.14 in the pre-peak region (the maximum load at the fifth step), the coefficients of interaction μ_b^{II} exhibit active interactions with the neighboring cracks, which facilitate the growth of crack B and thus the process of crack localization. On the other hand, the persistent small crack interactions at crack D as represented by μ_d^{II} in the postpeak regions provide sufficient clues as to why crack D cannot become the failure crack (of course due to the greater amount of fracture energy required to propagate several cracks in the nonlocalized cracking mode). Enlarging notch D to 50 mm in Case 6, crack D becomes the dominant crack after suppressing the simultaneous propagation of crack B at the peak load (Figure 5.13).

As shown in Figure 5.15 of Case 6, with little crack interactions involved in the propagation of crack D, the small coefficients of μ_d^{II} quickly vanish in the postpeak regions, indicating rapid crack localization. As expected, μ_b^{II} becomes large and persistent long after the peak load. Based on the obtained values of μ^{II}, it is observed that for beam problems with mixed-size notches, the coefficients of interaction of a large notch or crack are generally much smaller than the coefficients of interaction of a small notch or crack, except for the case when a dominant crack originates from this small notch, such as crack B in Case 5, where $\mu_d^{II} > \mu_b^{II}$ in the postpeak regions.

5.4 CRACK INTERACTIONS IN TUNNEL LININGS

To examine how cracks would interact with one another in different types of structures, the fracture test of the tunnel specimen discussed in Chapter 4 is reexamined to obtain the coefficients of interaction, using the half-EF-model shown in Figure 4.17. With caving above the ceiling area and uniform pressure loads applied to the sidewalls, the obtained cracking behaviors and the locations of the tension zone are shown in Figure 5.16, which presents numerical results on structural response and crack propagation charts.

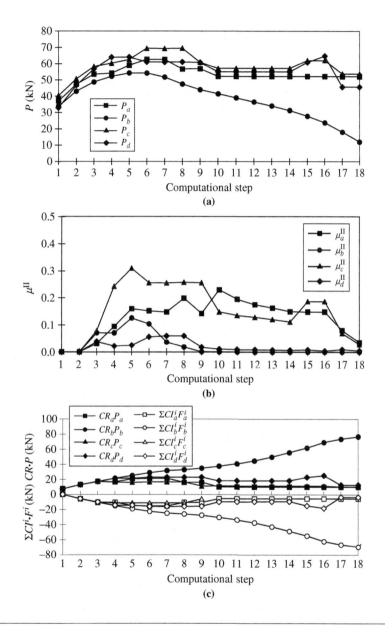

FIGURE 5.14 (a) External loads, (b) coefficients of interaction, and (c) principal tip force components for each assumed active crack (Case 5).

As shown in the deformed tunnel specimen, three tension zones exist: the outside of the ceiling area, the corner of the bottom plate, and the inside of the sidewall. Three cracks—A, B, and C—propagate in these tension zones from the initial notches that are preset in the numerical model with the same size of 25 mm, or one-tenth of the wall thickness. Unlike a beam under

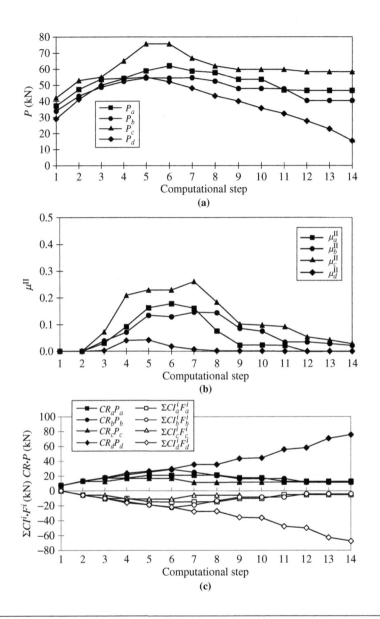

FIGURE 5.15 (a) External loads, (b) coefficients of interaction, and (c) principal tip force components for each assumed active crack (Case 6).

bending where all of the cracks originate from the same tension side, crack interactions in the tunnel specimen exhibit characteristics that are typical to structures with several tension zones.

As discussed previously, under the pressure loads, crack B in the sidewall was found to be the most active crack, followed by the growth of crack C in the bottom plate. The active propagation of crack C occurred in two separate stages: in the prepeak region and in the postpeak regions as

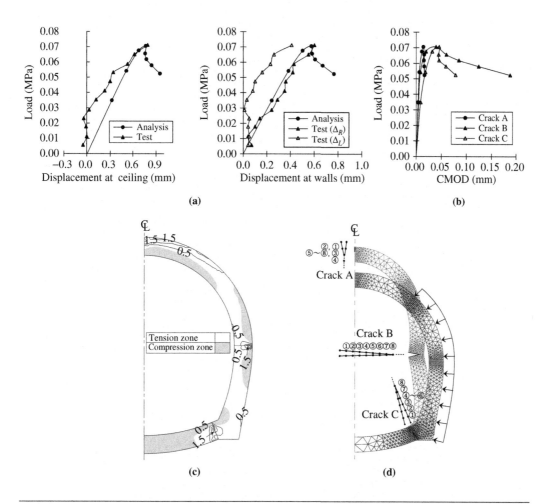

FIGURE 5.16 Numerical results on structural response and crack propagation (half-FE-model of tunnel specimen): (a) load-displacement relation, (b) load-CMOD relation, (c) maximum principal stress contours (MPa), and (d) crack propagation chart.

the crack became temporarily nonpropagating at around the peak load. Crack A in the ceiling area was the least-active crack, whose slow growth stopped completely after reaching the peak load. These cracking behaviors are the direct results of the deformation characteristics of the tunnel specimen under the given load and boundary conditions.

As shown in Figure 5.17, although the minimum required load for propagating a crack at steps 1 to 3 is found in P_c, the subsequent crack analysis results in the propagation of not only crack C but crack B as well. Taking into account the large structural deformation that takes place with each crack increment (set at one-tenth the wall thickness), it is obvious that the simultaneous growth of crack B with crack C is the result of correcting the single-active-crack mode because the tip force at crack B must have surpassed the tensile strength of concrete. As the structural deformation is gradually accumulated in the sidewall after the first three steps of crack propagation, P_b becomes the

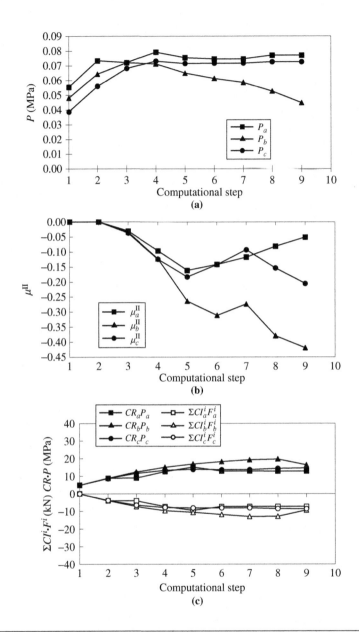

FIGURE 5.17 (a) External loads, (b) coefficients of interaction, and (c) principal tip force components for each assumed active crack (half-FE-model of tunnel specimen).

minimum required load to propagate crack B from the fourth step forward, while the growth of crack C is temporarily halted. As explained in the previous study of the full tunnel model in Chapter 4, the regrowth of crack C in the postpeak regions is directly related to the large structural deformations of the tunnel specimen due to the progressive opening of crack B in the sidewall, which cause a large bending moment to form at the bottom plate to reactivate crack C.

Perhaps the most interesting feature of the crack interaction in the tunnel specimen is represented by the negative coefficient of interaction, as shown in Figure 5.17. With several tension zones coexisting in the tunnel lining, the cohesive forces at one crack induce a compressive tip force component at another crack in a different tension zone that tends to close that crack. Apparently, this is the effect of reversing or resisting the structural deformation by the cohesive forces of a crack in any tension zones. Therefore, the crack interactions in these situations actually represent the resistance to the growth of a crack. This leads to an interesting phenomenon: The more active a crack becomes, the stronger resistance it encounters. This conclusion is drawn from the following facts in the μ^{II} diagram of Figure 5.17: $|\mu_b^{II}| > |\mu_c^{II}|$ from the fifth step as crack B becomes the sole leading crack, and $|\mu_c^{II}| > |\mu_a^{II}|$ from the eighth step as crack C grows again from the seventh step. Since crack A is the least active crack, relatively small coefficients of interaction $|\mu_a^{II}|$ are obtained.

Notice that as the coefficient of interaction becomes negative, the PTF coefficient μ^I must be greater than one, according to Eqs. (5.37) to (5.39). For instance, with $\mu_a^{II} = -0.16$, $\mu_b^{II} = -0.26$, and $\mu_c^{II} = -0.18$ obtained at the fifth step, the PTF coefficients become $\mu_a^I = 1.16$, $\mu_b^I = 1.26$, and $\mu_c^I = 1.18$, respectively. The stress concentration at crack B gradually becomes evident as the crack propagates in the postpeak regions, though with a much-reduced magnitude as compared to the scale of the stress concentration observed at a propagating crack in the beam problems discussed in the previous section.

5.5 CHARACTERISTICS OF CRACK INTERACTIONS WITH ONE AND MULTIPLE TENSION ZONES

The contrasting characteristics of the coefficients of interaction and the PTF coefficients of two cracks with one and two tension zones are compared in Figure 5.18. In the case of one tension zone, the two cracks interact directly through the same stress field as the cohesive forces of one crack induce a tensile force at the tip of the other and thus facilitate its propagation (in essence, due to the reduced structural stiffness and increased structural deformation), and vice versa. Therefore, the coefficients of interaction involving the two cracks are positive, and the PTF coefficients must be less than or equal to one, as shown in Figure 5.18a. On the other hand, if the deformed structural member possesses two separate tension zones and each contains a single crack, the two cracks interact not through the same tension field but through the general structural deformation that leads to the formation of the two tension zones, as shown in Figure 5.18b.

As the cohesive forces of one crack tend to close that crack and thus resist general structural deformation, a compressive force is induced at the tip of the other crack, which is equivalent to increasing the material resistance to fracture. Hence, the coefficients of interaction of the two cracks become negative, and the magnitudes of these coefficients represent the amount of the increased resistance encountered in propagating these cracks. Notice that this increment of resistance is reflected in the PTF coefficients that are greater than one, as shown in Figure 5.18b. Consequently, with multiple tension zones coexisting, the more active a crack becomes, the larger the structural deformation it causes, which in turn activates more cracks in other tension zones and thus results in more interaction or resistance to the propagation of that active crack. This explains clearly why a propagating crack encounters more resistance than a less active crack does.

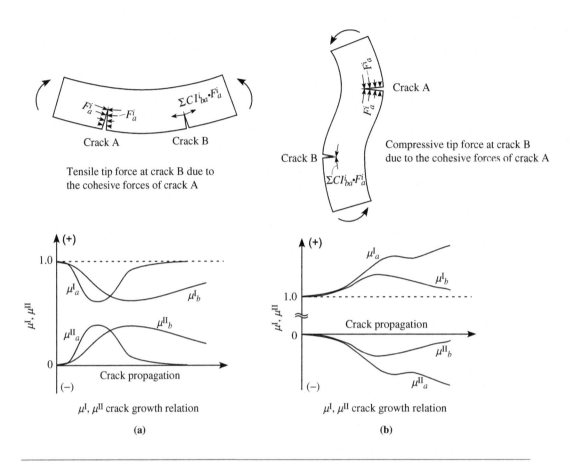

FIGURE 5.18 Characteristics of crack interactions with (a) one and (b) two tension zones.

If one of the tension zones contains several cracks, the cracks within that zone interact the same way as cracks in a single-tension zone problem, and the associated coefficients of interaction are positive. The coefficients of interaction involving cracks in other tension zones, however, must be negative because these interactions are tantamount to the material resistance encountered by a propagating crack, as just discussed.

As the PTF coefficient of a propagating crack represents the ratio of the tip force components due to the external load and the cohesive forces of the crack to the critical tip force for crack propagation, its inverse defines an influence factor that measures the magnifying or diminishing effect of the crack interactions on the propagation of that crack. This leads to

$$\beta = \frac{1}{\mu^{\mathrm{II}}} = \frac{1}{1 - \mu^{\mathrm{I}}} \tag{5.40}$$

Figure 5.19 shows the influence factors of Cases 1 to 6, in which the magnifying effect of the crack interactions on each of the main cracks is clearly visible in most of the cases with the

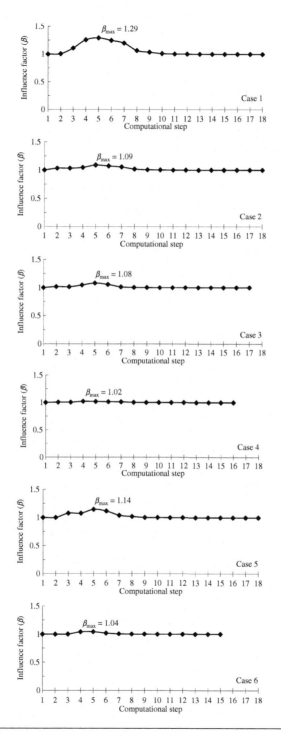

FIGURE 5.19 Influence factors of main cracks in Cases 1 to 6.

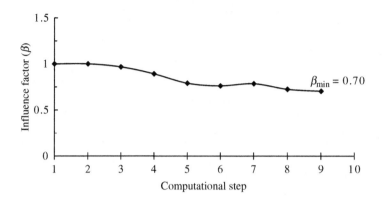

FIGURE 5.20 Influence factors of crack B in the half-FE-model of tunnel specimen.

maximum in Case 1 at 1.29, and the minimum in Case 4 at 1.02. On the other hand, Figure 5.20 illustrates the influence factors of the tunnel problem, which exhibits the diminishing effect of the crack interactions on crack B with the minimum of 0.70.

REFERENCES

Barpi, F. and Valente, S. (1998). "Size-effects induced bifurcation phenomena during multiple cohesive crack propagation." *J. Solids Struct.*, 35, 1851–1861.

Bazant, Z. P. and Wahab, A. B. (1980). "Stability of parallel cracks in solids reinforced by bars." *J. Solids Struct.*, 16, 97–105.

Carpinteri, A. and Monetto, I. (1999). "Snap-back analysis of fracture evolution in multi-cracked solids using boundary element method." *Int. J. Fracture*, 98, 225–241.

Ingraffea, A. R., Gerstle, W. H., Gergely, P., and Saouma, V. (1984). "Fracture mechanics of bond in reinforced concrete." *J. Struct. Eng.*, ASCE, 110(4), 871–890.

Kaplan, M. F. (1961). "Crack propagation and the fracture of concrete." *J. Am. Concrete Inst.*, 58(11), 591–610.

Murakami, Y. (1987). *Stress Intensity Factors Handbook*, Pergamon Press.

Shi, Z., Suzuki, M., and Ohtsu, M. (2004). "Discrete modeling of crack interaction and localization in concrete beams with multiple cracks." *J. Advanced Concrete Technology*, JCI, 2(1), 101–111.

Uchida, Y., Rokugo, K., and Koyanagi, W. (1993). "Flexural tests of concrete beams." In: JCI Com. Rep. *Applications of Fracture Mechanics to Concrete Structures*, pp. 346–349, Japan Concrete Institute.

Failure Modes and Maximum Loads of Notched Concrete Beams

6.1 INTRODUCTION

In the testing and analysis of various fatigue problems, the time-varying nature of cyclic loads on structures induced by winds, waves, vehicles, and so on is often subjected to oversimplification with loads of varying amplitudes but fixed loading positions. Experimental studies have shown that during fatigue tests of reinforced concrete beams and plates, the use of a moving cyclic load in a simulation of traffic loads may cause a reduction in the maximum load obtained under fixed-point monotonic or cyclic load conditions (Okada et al., 1982; Kawaguchi et al., 1987). Though these experimental observations have shed some light on the effects of changing loading positions during cyclic loading on the load-carrying capacity of a structure, the exact cause of the reduction is not well understood.

It is well known that fatigue mechanisms are complicated, and it has become increasingly clear that the phenomenon is closely related to multiple-crack activities during cyclic loading. This chapter begins with numerical studies on the load-carrying capacity of a notched concrete beam subjected to various monotonic loads, focusing on the change of cracking behavior and the maximum load as the load condition changes. It is confirmed that the cracking behavior and the failure mode are very sensitive to the loading positions, which could alter significantly the load-carrying capacity of a simple beam. Also, it is found that under a given load condition, increasing the number of initial notches in a numerical model may not necessarily reduce the flexural strength of the beam. Next, numerical studies using cyclic loads of varying loading position are carried out, aiming at clarifying the effects of changing loading positions during cyclic loading on cracking behavior and the load-carrying capacity of a notched beam.

The results of the preceding numerical analyses clearly demonstrate the close relationship between the failure mode and the load-carrying capacity of a notched beam. But exactly how the decrease of the load-carrying capacity of a structural member takes place during cyclic loading amid various cracking activities remains to be clarified. In a series of numerical studies in which the size of a specific notch was enlarged incrementally, the existence of a critical size at which the fracture process changed abruptly was found, and a significant reduction of the maximum load was obtained with the new failure mode (Shi and Suzuki, 2004).

The loading conditions and notch arrangements of the simple beams in the original study are illustrated in Figure 1.13. As shown, among the three notches introduced into the beam, A and B were kept at a constant size of 10 mm, while notch C was assigned various sizes to study the relations between the maximum loads and the failure modes under eccentric loading. The obtained relations are shown in Figure 1.14, which contains two curves. When the eccentric load was applied at notch C, a monotonically decreasing relation between the peak load and the size of notch C was obtained, and the dominating crack for beam failure was shown to invariably develop from notch C.

On the other hand, when the eccentric load was applied at notch A, the obtained maximum load seemed to be unaffected by the enlargement of notch C until it reached a critical value, beyond which the peak load decreased quickly as the size of notch C further increased. It was shown that a sudden change in the failure mode took place in the latter case. Before reaching the threshold value of notch C, the dominating crack for the beam failure originated from notch A; beyond that point it propagated from notch C. This important failure mechanism involving the sudden change of failure mode is the focus of study in this chapter.

To provide experimental evidence for the transition of the failure mode and the resulting reduction of the load-carrying capacity, an experimental study will be introduced (Shi et al., 2007). The tests focus on the maximum loads of notched concrete beams under four-point bending and the corresponding failure modes as the size of a control notch is gradually enlarged. Then numerical analyses are carried out to reproduce the fracture processes and obtain the maximum loads, which both compare well with the test results. These studies reconfirm the previous findings that for a given loading condition, the failure mode and the maximum load seem to be insensitive to the sizes of certain initial notches as long as they are less than a critical value. As the notch size surpasses that threshold value, however, a drastic change in the failure mode takes place, and the load-carrying capacity can drop significantly.

In studies of metal fatigue, it has long been known that the fatigue strength of a test specimen is not affected by introducing artificial micro holes into the specimen unless the size of these artificial defects exceeds a critical value, beyond which a significant degradation in fatigue strength takes place (Murakami, 1993). Related experimental studies will be briefly introduced, and the similarities between the two phenomena will be compared. Finally, engineering implications of the present study in clarifying the mechanisms of fatigue failure will be discussed.

6.2 NUMERICAL ANALYSIS OF NOTCHED BEAMS UNDER VARIOUS LOAD CONDITIONS

By assuming a variety of load conditions, this section presents crack analysis of notched beams, focusing on the maximum load and the failure mode.

6.2.1 Maximum Loads with Monotonic Loadings

This section presents crack analysis of notched beams under monotonic loadings.

Numerical Models

Figure 6.1 illustrates the round-robin fracture tests of unnotched plain concrete beams under four-point bending (Uchida et al., 1993) and FE models of notched beams subjected to various load conditions. As seen, initial notches are introduced in each numerical model in the three critical locations for crack propagation under the test condition—that is, notch A and notch C under

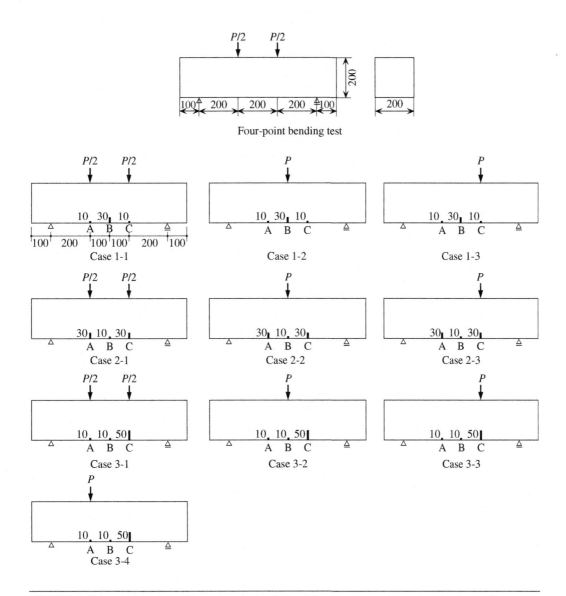

FIGURE 6.1 Numerical models (dimensions in mm) of notched beams under various load conditions.

the two loading points and notch B at the center of the beam. By changing the sizes of these initial notches, three types of the notched beam model are chosen to calculate the load-carrying capacity under each load.

In Cases 1-1 to 1-3, the two side-notches A and C are assumed to be small with the same size of 10 mm, while the central notch B with the size of 30 mm is three times larger. In contrast, in Cases 2-1 to 2-3, the side-notches are enlarged to 30 mm, and the central notch is kept at 10 mm.

As shown, the three load conditions considered include four-point bending, three-point bending, and a single load at the location above notch C. Apparently, whether the single load is applied at C or at A does not affect the load-carrying capacity because the notch arrangement is symmetrical with respect to the center of the beam. In Cases 3-1 to 3-4, notches A and B are both 10 mm, and notch C is assumed to be as large as 50 mm. Besides the three types of loadings just described, a single load is applied at A in Case 3-4.

To solve the crack equations, the bilinear tension-softening relation as shown in Figure 4.9 is employed. The material properties of the test specimen are summarized in Table 4.1.

Results and Discussion

Figures 6.2 to 6.4 summarize the numerical results, which include the load-displacement relations and the crack propagation charts. The maximum loads obtained under the prescribed load conditions are listed in Table 6.1 for each type of notched beam.

Cases 1-1 to 1-3   In view of the symmetrical feature of the loading conditions in Cases 1-1 and 1-2, the two problems are solved numerically using half-FE-models. With a single load at C in Case 1-3 the problem is asymmetric and, therefore, is analyzed using a full-model. As shown in Figure 6.2, with a larger central notch and two smaller side-notches, a single crack propagates at the center under the four- and three-point bending in Cases 1-1 and 1-2. In Case 1-3, although the fracture of the beam is also caused by a dominating central crack that is curvilinear with its tip tilting toward the loading point, a small crack also appears at notch C. As shown in the crack propagation charts, the peak load is reached at the sixth step in all the three cases, and the small crack at notch C in Case 1-3 is forced to close after the eighth step as the main crack penetrates beyond half of the beam depth.

As shown in Table 6.1, a significant difference between the peak loads in Cases 1-1 and 1-2 is found, which amounts to a reduction rate of 25 percent in terms of the load-carrying capacity of the simple beam as the loading is changed from the four-point bending to the three-point bending. Obviously, the smaller peak load in Case 1-2 is caused by the larger moment at the center of the beam generated under the three-point bending than under the four-point bending (as the former is 50 percent greater than the latter). Though the loading conditions in Cases 1-1 and 1-3 seem to be quite different, the peak load of Case 1-3 is almost identical to that of Case 1-1, which is largely due to the fact that the moments at the center of the beam from where the critical cracks propagate are exactly the same in both cases.

Cases 2-1 to 2-3   For the same reasons as in the previous cases, Cases 2-1 and 2-2 are analyzed using half-models, and the solution of Case 2-3 is based on a full-model, as shown in Figure 6.3. As the two side-notches are enlarged to 30 mm and the central notch is kept as small as 10 mm, the fracture of the beam does not initiate from the central notch but from the side-notches under all three loading conditions. In Cases 2-1 and 2-2, two cracks develop from notches A and C and propagate simultaneously until structural failure. Though a small crack does grow at notch B under the three-point bending in Case 2-2, it is forced to close as soon as the peak load is reached at the sixth step. In Case 2-3, notch C is the most critical location for crack propagation because the moment at notch C is twice as large as notch A, and the beam is fractured by a single crack there. As seen in the propagation charts, during that process a small crack also appears briefly at notch A.

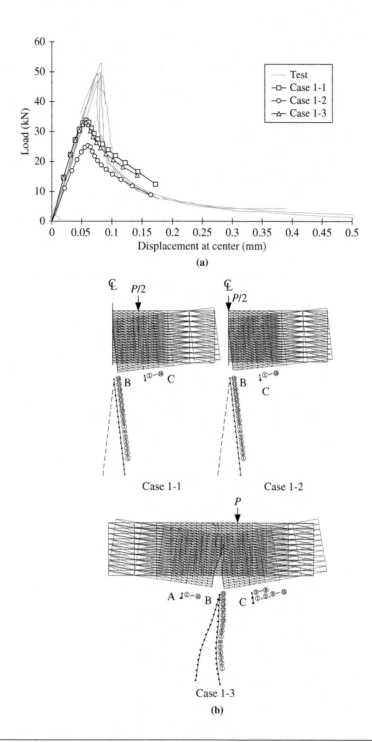

FIGURE 6.2 Numerical results of Cases 1-1 to 1-3: (a) load-displacement relation and (b) crack propagation charts (parentheses around numbers indicate maximum load).

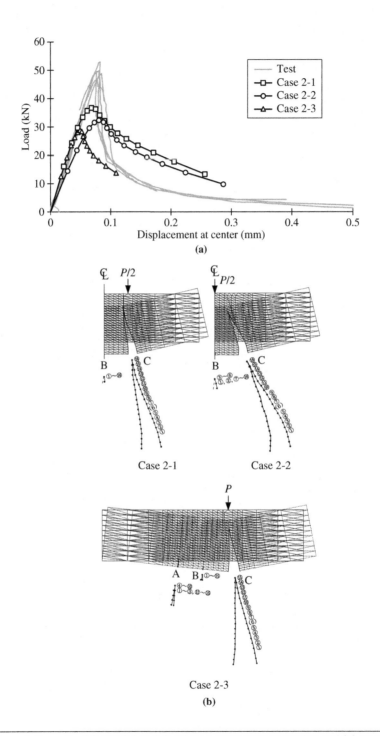

FIGURE 6.3 Numerical results of Cases 2-1 to 2-3: (a) load-displacement relation and (b) crack propagation charts (parentheses around numbers indicate maximum load).

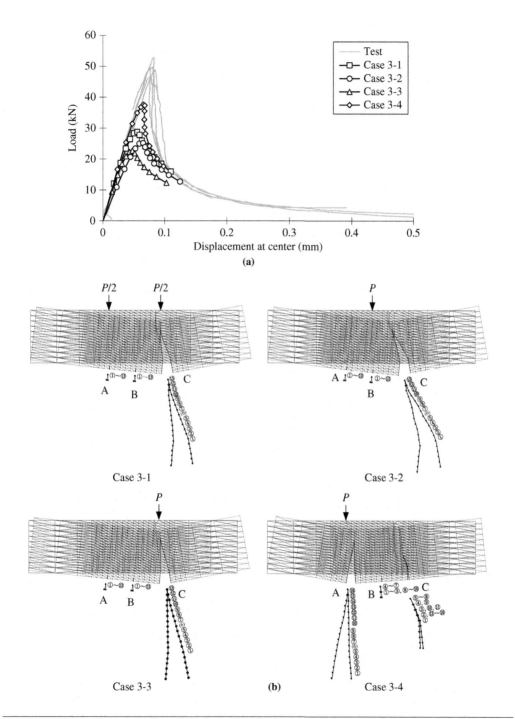

FIGURE 6.4 Numerical results of Cases 3-1 to 3-4: (a) load-displacement relation and (b) crack propagation charts (parentheses around numbers indicate maximum load).

Table 6.1 Maximum Loads Under Monotonic Loading (Cases 1 to 3)

Case No.	Case 1			Case 2			Case 3			
	1-1	1-2	1-3	2-1	2-2	2-3	3-1	3-2	3-3	3-4
P(kN)	33.8	25.3	33.1	36.7	32.4	28.8	28.8	25.2	22.7	37.7

As shown in Table 6.1, the maximum and the minimum peak loads in this group are obtained respectively in Cases 2-1 and 2-3, and the difference is equivalent to a reduction rate of 22 percent as the loading changes from the four-point bending to a single load at C. The smaller peak load in Case 2-3 is due to its localized cracking behavior. This is in view of the fact that in Case 2-3 the fracture of the beam is caused by a single crack, whereas in Cases 2-1 and 2-2 it is directly linked with two simultaneously propagating cracks.

Cases 3-1 to 3-4 As shown in Figure 6.4, all of the cases in this group are analyzed with full models. As the size of notch C is increased to 50 mm while notches A and B remain as small as 10 mm, the fracture of the beam invariably takes place at notch C in Cases 3-1 to 3-3. With a single load applied directly above notch A in Case 3-4, however, the stress concentration at notch A also becomes critical for crack propagation as two cracks propagate simultaneously from notch A and notch C. Upon reaching the peak load at the sixth step, crack localization begins at crack A as the crack becomes the only propagating crack in the postpeak regions, while crack C temporarily becomes arrested at around the peak load. As seen in the propagation charts, a small crack appears briefly at notch B from the fourth to the seventh steps, which is followed by the eventual closure of crack C after the eighth step.

As expected, the maximum load is obtained in Case 3-4, which is the only case involving multiple-crack propagations. As is evident from Table 6.1, the differences between the peak loads are more remarkable than those of the previous cases, and so are the reduction rates. When compared to the four-point bending, the reduction rate is 13 percent in Case 3-2 and 21 percent in Case 3-3—similar to the ratios of the previous cases. When compared to Case 3-4, the gap widens, and the reduction rate jumps to 24 percent in Case 3-1, 33 percent in Case 3-2, and 40 percent in Case 3-3. As such, the load-carrying capacity of a notched beam may vary significantly depending on the details of a given load condition such as the loading position because cracking behavior is very sensitive to such details. It should be noted that the load-carrying capacity is a measure of the strength of resistance to the material failure, which becomes evident only through various fracturing processes.

6.2.2 Maximum Load Increase with Higher Density of Initial Notches

This section presents crack analysis of notched beams with higher density of initial notches.

Numerical Models

Figure 6.5 illustrates four numerical models that are derived from Cases 2-1 to 2-3 by increasing the density of initial notches to study its effect on the maximum load. As seen, notch D and notch E are introduced in between notches A and B and notches B and C with equal intervals,

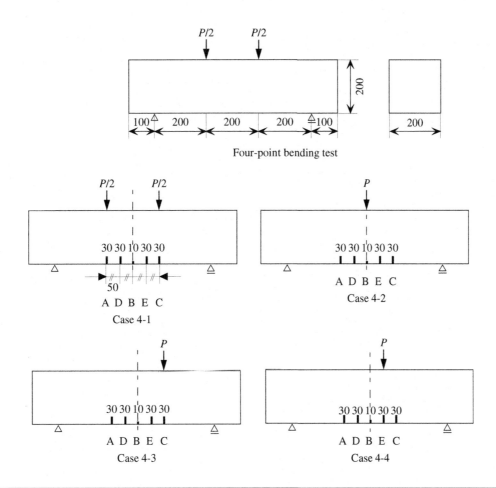

FIGURE 6.5 Numerical models (dimensions in mm) of notched beams with higher density of initial notches.

respectively. As with notches A and C, the new notches are also assumed to be 30 mm. Besides the three loading conditions studied in Cases 2-1 to 2-3, a single load is applied at E in Case 4-4. For the geometric details, refer to Figure 6.5.

Results and Discussion

The results of crack analysis for Cases 4-1 to 4-4 are presented and discussed.

Cases 4-1 to 4-4 For the same reasons as in the previous cases, Cases 4-1 and 4-2 are analyzed using half-models, and the solutions of Cases 4-3 and 4-4 are based on full-models, as shown in Figure 6.6. As seen, with the increasing number of initial notches, the critical locations for crack propagation under the four-point and the three-point bending shift to notches D and E, alleviating the stress concentrations at notches A and C. This is manifested by the less-aggressive crack propagations at notches A and C that are confined only to the prepeak regions. In Cases 4-3 and 4-4 where a single load is applied

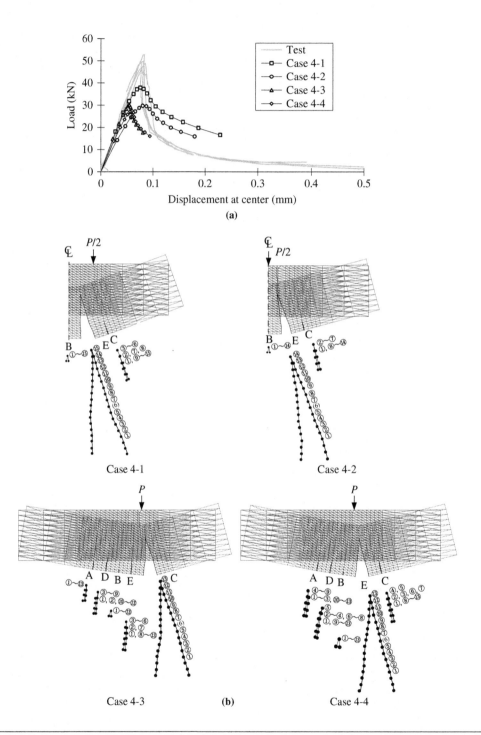

FIGURE 6.6 Numerical results of Cases 4-1 to 4-4: (a) load-displacement relation and (b) crack propagation charts (parentheses around numbers indicate maximum load).

Table 6.2 Maximum Loads Under Monotonic Loading (Case 4)

Case No.	Case 4			
	4-1	4-2	4-3	4-4
$P(kN)$	37.9	29.7	30.0	27.4

eccentrically, the fracture of the beam is caused by a single crack propagating beneath the loading point, while small cracks appear at some of the other notches in the prepeak regions.

As shown in Table 6.2, the maximum and the minimum peak loads are obtained respectively in Cases 4-1 and 4-4, and the reduction rate in the latter case approaches 30 percent. It is interesting to note that despite the apparent differences found in Cases 4-2 and 4-3 regarding the loading conditions and cracking behaviors, the two peak loads are almost identical with a reduction of more than 20 percent from Case 4-1. Comparing Cases 4-1 and 4-2 and Cases 4-3 and 4-4 based on the similarities of the cracking behaviors, it seems that the smaller peak load in each of the two cases is caused by the larger moment at the location of the critical crack or cracks generated under the corresponding load conditions (since the moment at E in Case 4-2 is 25 percent greater than that in Case 4-1, and the moment at E in Case 4-4 is 9 percent greater than that at C in Case 4-3).

A careful comparison of the peak loads between Cases 2-1 and 4-1, and Cases 2-3 and 4-3 in Tables 6.1 and 6.2 reveals an interesting fact: Under a given loading condition, increasing the number of initial notches may sometimes result in a larger peak load. In Case 4-1, the peak load is 3 percent higher than in Case 2-1 and is 4 percent higher in Case 4-3 than in Case 2-3 as the notch density increases. On the other hand, the peak load of Case 4-2 is reduced by 8 percent from that of Case 2-2 as a result of increasing the number of initial notches. It is obvious that the increase or decrease of the peak loads in these cases is due to the variations of the cracking behavior and the fracture process as the condition on initial notches varies, as shown in the crack propagation charts in Figures 6.3 and 6.6.

Regarding the strength increase in Cases 4-1 and 4-3, a similar phenomenon has been reported in fatigue tests of concrete beams in which the flexural strength of a concrete beam after a limited number of cyclic loadings was found to be greater than that obtained under only monotonic loading (CAJ, 1985). This seemingly paradoxical phenomenon can be explained by the fact that during the cyclic loading, a certain degree of material damage must have been inflicted upon the beam either microscopically or macroscopically, a situation comparable to the increasing density of initial notches in Cases 4-1 and 4-3. Thus, when monotonic loading is subsequently imposed, a different cracking process may emerge and lead to a strength increase, as observed in the present numerical cases.

6.2.3 Maximum Loads with Alternative Loadings

This section presents crack analysis of notched beams with alternative loadings.

Numerical Models

As has been demonstrated through numerical studies using monotonic loads, the load-carrying capacity of a notched beam is very sensitive to the loading positions. In this section, the fractures of the simple beams under four-point bending in Cases 2-1 and 4-1 are studied using cyclic loads of varying loading position in Cases 5-1 to 5-3. As shown in Figure 6.7, a single load is applied

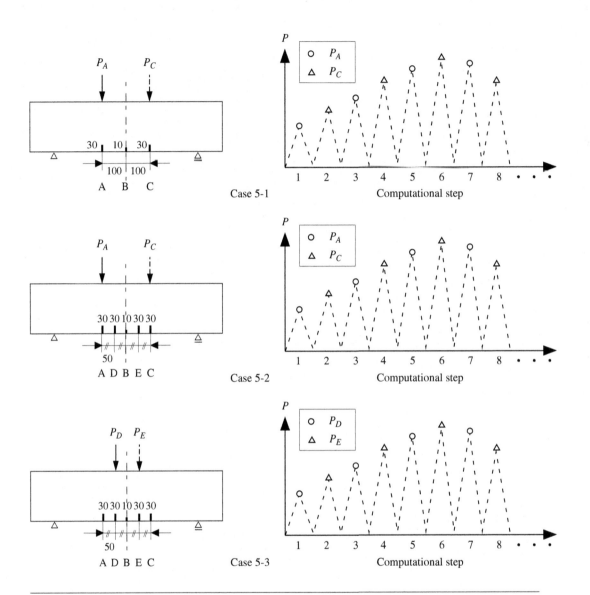

FIGURE 6.7 Numerical models (dimensions in mm) of notched beams with alternative loadings.

alternatively at A and C in Cases 5-1 and 5-2, and at D and E in Case 5-3, with the objective of inflicting similar material damage to the beam as with the four-point bending but through different cracking processes.

In the crack-tip-controlled numerical analysis, the peak load in each cycle of loading is obtained by solving the crack equations for the external load required for propagating an active crack to the next nodal point along the preset crack path. Stress analysis is then carried out at these peak loads in terms of the total stress and strain. This is different from the incremental stress analysis employed for monotonic loading because the stress fields obtained at any two consecutive peak loads are considered to be independent as a result of changing loading positions. Furthermore, the crack paths are prescribed as vertical to facilitate the convergence of numerical solutions. Except for these two simplifications, crack analysis is carried out following the same numerical procedures as with monotonic loads.

Results and Discussion

The results of crack analysis for Cases 5-1 to 5-3 are presented and discussed.

Cases 5-1 to 5-3   Figure 6.8 presents numerical results of Case 5-1, which include the peak loads P_A and P_C of the cyclic loading at each step, the load-displacement relations, and the crack propagation charts. As seen, the fracture of the beam is caused by two propagating cracks from notches A and C, which closely resembles the structural damage sustained under the four-point bending in Case 2-1, and the maximum load is reached at the tenth step. By shifting the loading position between A and C during the cyclic loading, however, the maximum load drops to $P_{max} = 28.9$ kN, causing a 21 percent reduction from $P_{max} = 36.7$ kN of Case 2-1. The cause of this strength reduction lies in the differences between the two fracture processes.

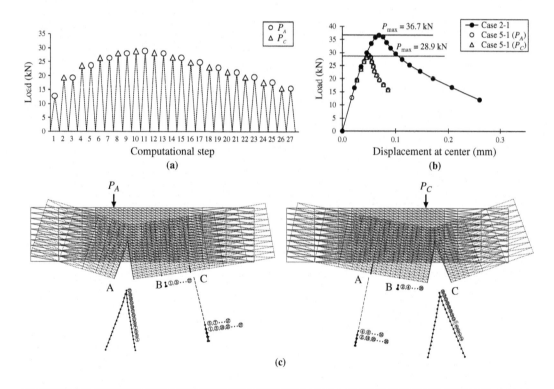

FIGURE 6.8 Numerical results of Case 5-1: (a) cyclic load at each step, (b) load-displacement relation, and (c) crack propagation charts (parentheses around numbers indicate maximum load).

As clearly demonstrated in the propagation charts, the simple mode of the simultaneous propagations of the two cracks under the four-point bending in Figure 6.3 is replaced by a fracture process that involves two localized cracking modes exchanging after each cycle of loading. As the loading position varies, the crack under the acting load is activated and extends from its tip by one mesh-size of 10 mm, while the other is closed due to the variations of the stress fields. It should be pointed out that the propagation of an active crack from its very tip immediately after its previous closure relies on the damage accumulation along the path of each crack. Similar to Case 2-3, the localized cracking modes under the cyclic loading are considered to be the main cause of the decrease in the maximum load.

The numerical results of Cases 5-2 and 5-3 are given in Figures 6.9 and 6.10, respectively. As shown, the fracture of the beam is caused by two propagating cracks from notches A and C in Case 5-2 and from D and E in Case 5-3. The structural damage in Case 5-3 is analogous to that in Case 4-1. Compared to $P_{max} = 37.9$ kN in Case 4-1, the maximum load is decreased to $P_{max} = 29.9$ kN in Case 5-2 at the eleventh step with a reduction of 21 percent, and to $P_{max} = 27.5$ kN in Case 5-3 at the tenth step with a reduction as high as 27 percent. Obviously, the decrease of the flexural strength in these cases is attributed to the localized cracking processes observed in the

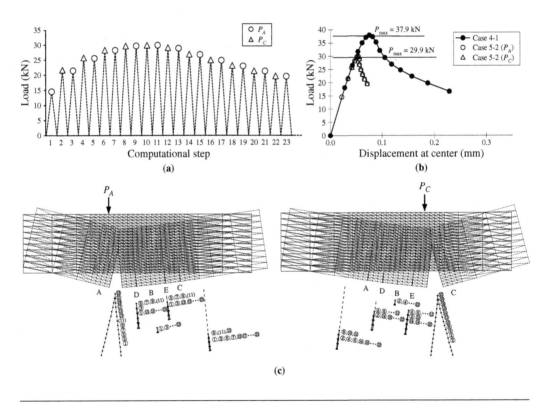

FIGURE 6.9 Numerical results of Case 5-2: (a) cyclic load at each step, (b) load-displacement relation, and (c) crack propagation charts (parentheses around numbers indicate maximum load).

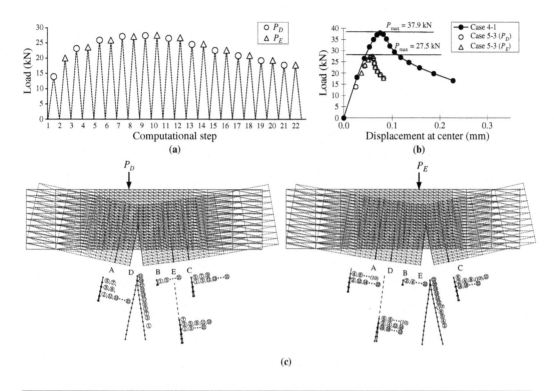

FIGURE 6.10 Numerical results of Case 5-3: (a) cyclic load at each step, (b) load-displacement relation, and (c) crack propagation charts (parentheses around numbers indicate maximum load).

propagation charts of Figures 6.9 and 6.10. With its reduction approaching 30 percent, Case 5-3 highlights the possibility of a significant strength loss during cyclic loading as the number of initial notches increases.

As such, the load-carrying capacity of a notched beam depends very much on the kind and concentration of initial notches as well as the resulting fracture processes. During cyclic loading, the damage accumulation plays an important role in the crack propagation as in Cases 5-1 to 5-3, and the reduction of the load-carrying capacity is actually the result of changing fracture processes due to the activation of the critical failure mode or modes.

6.3 CRITICAL INITIAL NOTCH AND ITS INFLUENCE ON FAILURE MODE AND THE MAXIMUM LOAD

This section presents crack analysis of notched beams with various notch sizes, focusing on the critical initial notch and its influence on failure mode and the maximum load.

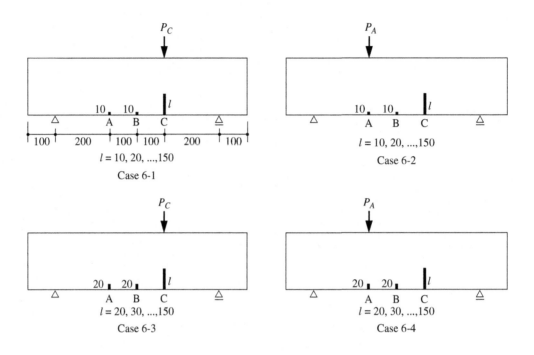

FIGURE 6.11 Numerical models (dimensions in mm) of notched beams with eccentric loading to obtain relations of failure mode and the maximum load.

Numerical Models

To provide a mechanism for the reduction of the load-carrying capacity during cyclic loading, two sets of numerical cases are studied in this section, as shown in Figure 6.11. Obviously, these cases are inspired by the study of Cases 3-3 and 3-4. As seen, Cases 6-1 and 6-2 are the extensions of Cases 3-3 and 3-4 by assigning various sizes to notch C, and Cases 6-3 and 6-4 are the variations of Cases 6-1 and 6-2 by enlarging the size of notches A and B to 20 mm.

Results and Discussion

The results of crack analysis for Cases 6-1 to 6-4 are presented and discussed.

Cases 6-1 and 6-2 Numerically obtained relations between the maximum load and the size of notch C are presented in Figure 6.12. As the ratio of notch size to beam depth increases, the peak load P_C in Case 6-1 decreases monotonically. Clearly, with the load being applied directly above notch C, there is no uncertainty left in regard to the failure mode, and the fracture of the beam always occurs at notch C. As the load-carrying capacity is basically derived from the remaining ligament length at notch C, it is obvious that the larger the notch is, the smaller the load-carrying capacity will be. On the other hand, in Case 6-2, the peak load P_A remains approximately at 37.5 kN and seems to be unaffected by the very existence of notch C until the size of the notch

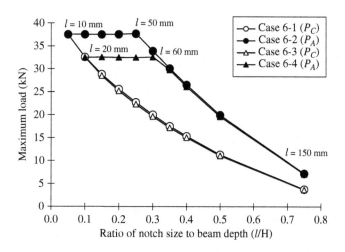

FIGURE 6.12 Relations between the maximum load and the ratio of notch size to beam depth (Cases 6-1 to 6-4).

exceeds 50 mm. As notch C increases beyond 50 mm, P_A decreases quickly at a faster rate than that of P_C in Case 6-1.

Numerical results show that a transition in the fracture process takes place at this stage. Up to the size of 50 mm, the fracture of the beam occurs at notch A; beyond that critical point, the beam fails at notch C. The direct involvement of two independent failure modes in the beam failure is the source of the decrease of the load-carrying capacity in Case 6-2. Obviously, the obtained invariant peak load in the region up to the size of 50 mm is due to the fact that the remaining ligament length at notch A from which the load-carrying capacity is derived is constant. When the change of failure mode takes place, however, an abrupt decrease of the peak load is inevitable because the load-carrying capacity is now dependent on the much smaller ligament length at notch C.

These numerical results indicate that for any given load condition, there may exist several potential failure modes in a structural member, and each of them is associated with a different fracture mechanism and thus may lead to a very different load-carrying capacity. The change of failure mode may occur as a result of changing the initial condition, such as by varying the size of notch C in the present study, or during cyclic loading (with varying loading position) as certain cracks propagate to some critical lengths to trigger other failure modes. As a result, the load-carrying capacity of the structural member can be greatly reduced.

The reduction rates of P_C to P_A are calculated and presented in Figure 6.13. With an extremely large notch of 150 mm at C (three-fourths of the beam depth), the ratio reaches approximately 50 percent. In other words, the flexural strength of the beam predicted by P_A could be twice as large as that of P_C. It is noted that P_A and P_C should stand for the same peak load if the beam is devoid of any initial defects.

Cases 6-3 and 6-4 In Case 6-4, upon increasing the size of notches A and B to 20 mm, the invariant peak load P_A is obtained approximately at 32.5 kN as notch C increases from 20 mm to 60 mm, which amounts to a decrease of 13 percent from the level of 37.5 kN in Case 6-2.

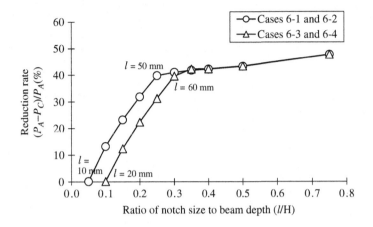

FIGURE 6.13 Relations between the reduction rate and the ratio of notch size to beam depth (Cases 6-1 to 6-4).

Again, two failure modes are involved: the notch A–type fracture or the notch C–type fracture. Apparently, the threshold value dividing the two is 60 mm; beyond that point, the beam fails at notch C. The corresponding results of Cases 6-3 and 6-4 are also shown in Figures 6.12 and 6.13, and the previous discussions on Cases 6-1 and 6-2 are basically valid for the present cases.

6.4 EXPERIMENTAL VERIFICATIONS ON RELATIONSHIPS BETWEEN FAILURE MODES AND THE MAXIMUM LOADS

This section presents experimental and numerical studies on notched beams under four-point bending, clarifying the relationship between failure modes and the maximum loads.

6.4.1 Four-Point Bending Tests

As is well known, the only sure means to verify a theory is through experiments. In this section, the experimental evidence for the strong dependence of the load-carrying capacity of a structural member on the failure mode is presented. The experimental investigation focuses on the maximum loads of notched concrete beams under four-point bending and the corresponding failure modes. The four-point bending test and geometric details of the specimens are shown in Photo 6.1 and Figure 6.14, respectively. Three small notches of the same size of 10 mm were introduced into plain concrete beams that were placed directly below the two loading points and at the midspan, respectively. Then a fourth notch was introduced into the same specimens, which varied in size and served as a control notch for switching the failure mode to demonstrate the transition of fracture processes in the beam at a critical notch size.

The mix proportions of concrete are summarized in Table 6.3, and the material properties of the test specimens are given in Table 6.4. Table 6.5 shows the test cases and notch arrangements.

PHOTO 6.1 Four-point bending test on relations of failure mode and the maximum load.

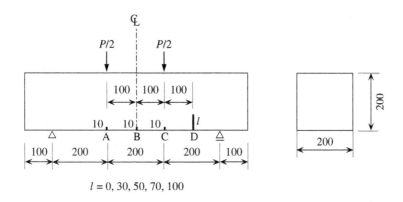

$l = 0, 30, 50, 70, 100$

FIGURE 6.14 Four-point bending tests (dimensions in mm) of notched concrete beams on relations of failure mode and the maximum load.

Table 6.3 Composition of Concrete Used in Fracture Tests

Maximum Size of Coarse Aggregate (mm)	Slump (cm)	Air (%)	W/C Ratio (%)	Sand-Coarse Aggregate Ratio (%)	Unit Weight (kg/m³)					
					Water W	Cement C	Fine Aggregate S	Coarse Aggregate G	Admixture	
									AE₁	AE₂
15	12.5	4.8	60.0	50.0	170	283	898	926	250 ml/C C = 100kg	C × 0.0005

Note: AE_1 = Air-entraining and water-reducing agent
AE_2 = Air-entraining agent

Table 6.4 Material Properties of Concrete Used in Fracture Tests

Material Property / Curing Period	Compressive Strength (N/mm^2)		Modulus of Elasticity (kN/mm^2)		Tensile Strength (N/mm^2)	
	Individual Specimen	Average	Individual Specimen	Average	Individual Specimen	Average
7 days	23.68	23.21	24.43	23.83	2.68	2.42
	22.66		24.06		2.15	
	23.30		22.99		2.44	
28 days	35.91	36.80	25.72	26.06	3.06	2.82
	39.22		27.16		2.41	
	35.27		25.30		2.98	

Table 6.5 Test Cases and Notch Arrangements of Fracture Tests

Test Case	Notch Size (mm)				Number of Specimens
	A	B	C	D	
1	10	10	10	0	3
2	10	10	10	30	3
3	10	10	10	50	3
4	10	10	10	70	3
5	10	10	10	100	3

As seen, five cases were tested with the size of notch D changing from 0 to 100 mm, and for each case three specimens were prepared.

The results of the tests are summarized in Table 6.6, which includes the maximum load and the location of the failure cross section. Samples of the failed beams are shown in Photo 6.2. With only three notches in Case 1, two specimens failed at notch B, and one broke at notch C, reflecting the influence of the inherent material variations on cracking behavior. By introducing an additional notch of 30 mm into the beam in Case 2, the three specimens all failed at notch B. The results suggest that notch D replaced notches A and C as a critical stress concentration point for potentially active cracks to emerge and to compete with cracks from notch B. In Case 3 one specimen broke at notch D, and two others failed at notch B, indicating that at 50 mm, notch D may have approached its threshold value for the failure mode to change. For Cases 4 and 5, all specimens failed at notch D, and their maximum loads were much lower. As seen, notch D indeed served as a control notch for altering

Table 6.6 Results of Fracture Tests

| Test Case | Specimen No. | Maximum Load (kN) | | Location of Fracture |
		Individual Specimen	Average	
1	1	50.0	48.8	Notch B
	2	50.6		Notch C
	3	45.7		Notch B
2	1	43.0	44.7	Notch B
	2	46.8		Notch B
	3	44.2		Notch B
3	1	44.7	43.1	Notch D
	2	38.4		Notch B
	3	46.2		Notch B
4	1	36.0	32.7	Notch D
	2	32.2		Notch D
	3	29.8		Notch D
5	1	22.2	21.6	Notch D
	2	21.8		Notch D
	3	20.8		Notch D

failure modes, and the experimental results clearly demonstrate the strong relationship between the load-carrying capacity of a structural member and the failure mode. Further discussions on these results will be presented in the following section, based on numerical analyses.

6.4.2 Numerical Analyses

In the following, crack analysis of the mode-I type is carried out to obtain the failure mode and the maximum load of the notched beam under the four-point bending. The FE mesh is illustrated in Figure 6.15, where the size of notch D is assigned values from 0 to 150 mm, with an increment of 10 mm up to the size of 80 mm. The material properties for the numerical study are summarized in Table 6.7.

The obtained analytical relations between the peak load and the ratio of notch size (notch D) to beam depth are shown in Figure 6.16, and the crack propagation charts are illustrated in Figure 6.17. Based on the analyses, the load-carrying capacity of the notched beam under the four-point bending remains approximately unchanged at 43.4 kN, as notch D increases from 0 mm to 50 mm; beyond that point, the peak load decreases quickly, with a reduction rate of more than 10 percent for every 10 mm enlargement of notch D.

Also shown in Figure 6.16 are the test results that clearly confirm the numerical predictions on the relations between the maximum load and the size of notch D, especially regarding the abrupt decrease of the maximum load beyond the critical notch size of 50 mm and the reduction rate.

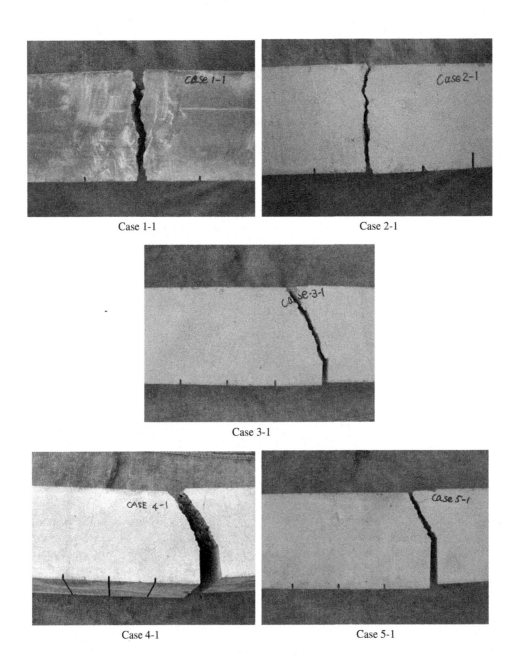

Case 1-1

Case 2-1

Case 3-1

Case 4-1

Case 5-1

PHOTO 6.2 Locations of the failure cross sections in fracture tests on relations of failure mode and the maximum load.

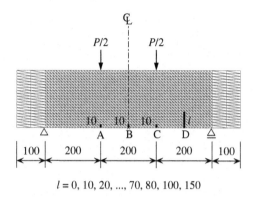

$l = 0, 10, 20, ..., 70, 80, 100, 150$

FIGURE 6.15 FE models of fracture tests (dimensions in mm) on relations of failure mode and the maximum load.

Table 6.7 Material Properties for Numerical Studies of Fracture Tests

E (kN/mm^2)	ν	f_c (N/mm^2)	f_t (N/mm^2)	G_F (N/mm)
26.1	0.2	36.8	2.8	0.1

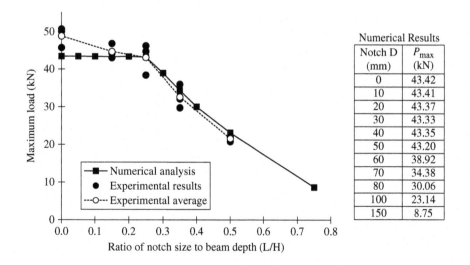

FIGURE 6.16 Relations between the maximum load and the ratio of notch size to beam depth.

As shown in Case 1 of the test, in which notch D was not yet introduced into the beam, two of the three tests recorded higher peak loads at the 50 kN level, which were 15 percent higher than the numerical prediction and are deemed to be within the normal range of variations for fracture tests of plain concrete beams.

As such, the numerical analysis predicts that the maximum load of the beam will not be affected by introducing notch D into the beam as long as the size of the notch does not exceed 50 mm. As shown in Figures 6.17A and 6.17B, the failure mode switches from the midspan fracture to the fracture at notch D, as the control notch reaches 50 mm or beyond. The obtained threshold value of 50 mm for the change of failure mode is deemed to be in close agreement with the experimental observations. Even though only one of the three test specimens failed at notch D during the experiments, the very fact that the failure modes began to diversify at the size of 50 mm indicates that a transitional point for transforming the fracture mechanism was reached.

Based on this observation, it seems proper to define a critical notch as a transitional point at which a new failure mode replaces the previous one while the maximum load basically remains unchanged. In other words, in the threshold region, two failure modes coexist, and both can lead to approximately the same maximum load. Obviously, the most important feature of the beam with the change of failure mode beyond the threshold region is the significant reduction in its load-carrying capacity, which is due to the decreasing ligament length at notch D. For reference, the analytical and experimental relations between load and CMOD at notch D are shown in Figure 6.18 for Cases 2 to 5. Notice that during the tests, data was recorded up to the peak load only.

6.5 ENGINEERING IMPLICATIONS

As is clearly demonstrated by the present study, when multiple cracks are involved in the fracture process of a structural member, the change of cracking behavior and the failure mode under certain circumstances may lead to a substantial reduction in the load-carrying capacity of that member. This finding may have significant implications in clarifying the fatigue mechanisms of certain engineering materials.

In studies of metal fatigue, it has long been known that the fatigue strength of a test specimen is not affected by introducing artificial micro holes into the specimen unless the size of these artificial defects exceeds a critical value, beyond which a significant reduction in fatigue strength takes place (Murakami, 1993). Figure 6.19 illustrates two types of carbon steel specimens and the shape of 12 identical micro holes, which were introduced into these specimens at three locations (A, B, and C) with the equal spacing of 5 mm. At each location, along the perimeter of a cylindrical specimen, four micro holes were drilled into the specimen at equal intervals.

Details on the material properties of the test specimens and test procedures can be found elsewhere (Murakami, 1993). Relations between the experimentally obtained fatigue limit (the stress level below which a test specimen will never break no matter how many stress cycles are applied) and the size of micro holes under rotating bending are shown in Figure 6.20. As seen, a critical diameter for these micro holes exists for each type of material, which is approximately 70 μm

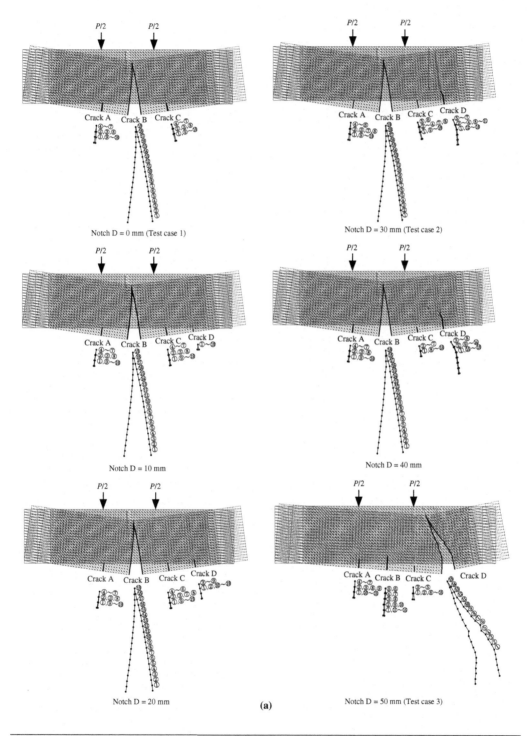

FIGURE 6.17A Crack propagation charts of fracture tests on relations of failure mode and the maximum load.

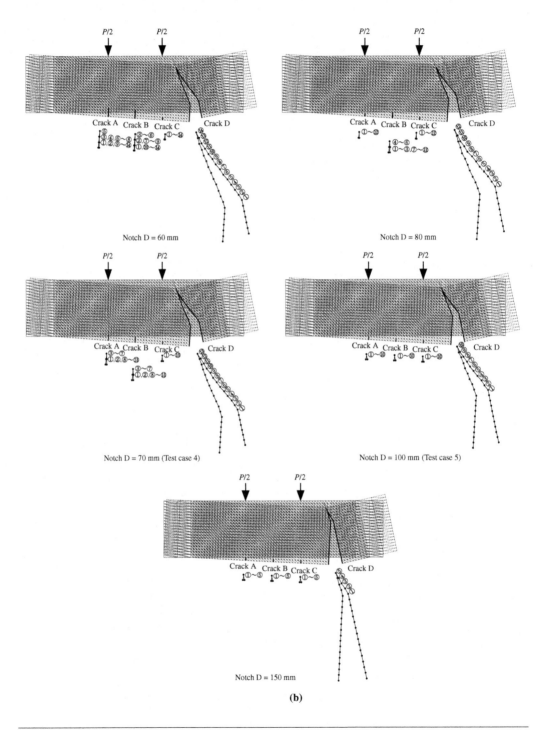

FIGURE 6.17B Crack propagation charts of fracture tests on relations of failure mode and the maximum load.

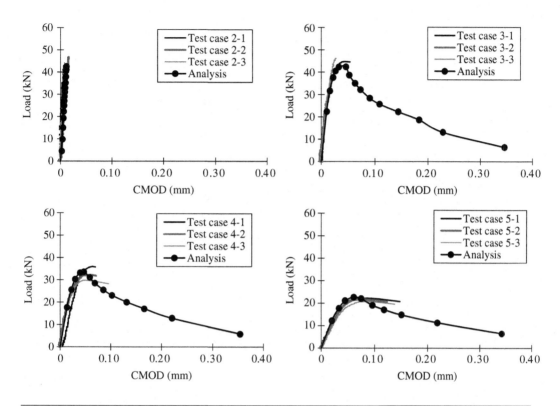

FIGURE 6.18 Load-CMOD relations at notch D of fracture tests on relations of failure mode and the maximum load.

for S10C steel and 35 μm for S45C steel. By introducing micro holes that are smaller than the critical size into a specimen, no particular influence on fatigue strength can be observed, and the fatigue limit for smooth specimens (without micro holes) is obtained; otherwise, a reduction in the fatigue limit occurs. It is known that the critical sizes of 70 μm and 35 μm are closely related to the fact that the maximum nonpropagating crack observed at the fatigue limit in annealed, smooth specimens is approximately 100 μm for S10C steel and 50 μm for S45C steel under rotating bending.

As is observed in Figures 6.16 and 6.20, a strong similarity exists in the strength-defect size relations obtained from the two strength tests, although these tests were carried out on different materials (concrete versus steel) and under different loading conditions (monotonic loading versus cyclic loading). Because the ultimate strengths of concrete and steel are all governed by fracturing processes that involve crack initiation and propagation (based on different material rules, of course) no matter whether under monotonic or cyclic loading, the fundamental mechanisms for strength degradation in the two situations may well be closely related.

Based on the circumstantial evidence shown in Figure 6.20, the degradation of fatigue strength beyond the threshold region may indicate fundamental changes in the cracking behavior and the

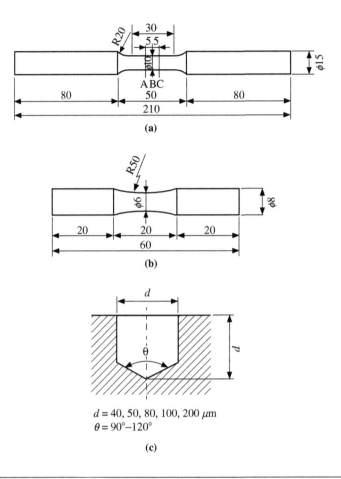

FIGURE 6.19 Test specimens (dimensions in mm) and the shape of micro holes from fatigue tests: (a) S10C specimen, (b) S45C specimen, and (c) shape of micro hole (after Murakami, 1993; courtesy of Yokendo).

failure mode in test specimens having larger micro holes, as is inferred from the study of the strength degradation of notched concrete beams. More specifically, different fracture mechanisms may have been involved in the fatigue tests just discussed: one that governs the fatigue strength of the smooth test specimens and one that is activated only after introducing micro holes whose sizes surpass the critical value. Perhaps it is the latter that causes the reduction in the fatigue strength of the test specimens for S10C and S45C steels.

In general, initial flaws and defects exist in all engineering materials and thus in all structural members; only the degree of imperfection varies. Under actual cyclic loading, not only the amplitude but also the loading position may change. Eventually, the material weakening process of a

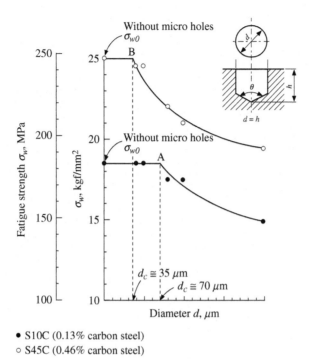

FIGURE 6.20 Relations between fatigue limit and size of micro holes (after Murakami, 1993; courtesy of Yokendo).

structural member caused by repeated loading inevitably involves multiple cracking originated from some of these spatially distributed initial imperfections, and diversified cracking behaviors can be expected. As a threshold value in terms of a critical crack length or a critical crack density (such as the maximum number of cracks in a certain location) and so forth is approached, new cracking behaviors may abruptly emerge and replace the previous ones, causing strength degradation. Obviously, this process can repeat itself until the remaining material strength can no longer sustain the level of stress produced by the design load, leading to abrupt structural failure.

To illustrate the point, Figure 6.21 presents numerical and experimental results on the maximum load of a notched beam by arbitrarily increasing the sizes of the initial notches to obtain a multistage strength degradation curve. Based on this analysis, the degradation of fatigue strength in fatigue tests of reinforced concrete beams and plates introduced in the beginning of this chapter, in which a moving cyclic load was employed, can be readily understood.

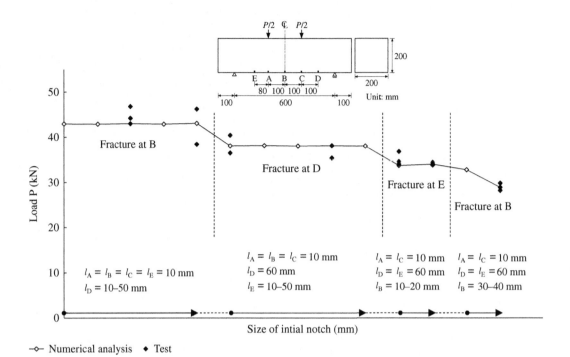

FIGURE 6.21 Numerical and experimental results on multistage strength degradation due to change of failure mode in bending tests of notched concrete beams.

REFERENCES

Cement Association of Japan (CAJ). (1985). "Study on the flexural fatigue strength of concrete." *Report R-3*, Committee on Cement-Concrete Road, Tokyo, 3–27.

Kawaguchi, M., Kawaguchi, T., Harada, K., and Takahashi, M. (1987). "Fatigue tests of reinforced concrete slab models of a highway bridge and an attempt to diagnose their residual lives." *Proc. of JSCE*, 380(I-7), 283–292.

Murakami, Y. (1993). "Effects of shape and size of micro defects on fatigue limit." *Metal Fatigue: Effects of Small Defects and Nonmetallic Inclusions*, Yokendo, 33–53.

Okada, K., Okamura, H., Sonoda, K., and Shimada, I. (1982). "Cracking and fatigue behavior of bridge deck RC slabs." *Proc. of JSCE*, 321(5), 49–61.

Shi, Z. and Suzuki, M. (2004). "Numerical studies on load-carrying capacities of notched concrete beams subjected to various concentrated loads." *Construction and Building Materials*, 18, 173–180.

Shi, Z., Nakano, M., and Yamakawa, K. (2007). "Experimental and numerical studies on the load-carrying capacity of notched concrete beams." *New Trends in Fracture Mechanics of Concrete*, A. Carpinteri, P. Gambarova, G. Ferro, and G. Plizzari eds., Taylor & Francis, pp. 285–291.

Uchida, Y., Rokugo, K., and Koyanagi, W. (1993). "Flexural tests of concrete beams." *Applications of Fracture Mechanics to Concrete Structures, JCI Com. Rep.*, Japan Concrete Institute, 346–349.

Mixed-Mode Fracture

7

7.1 INTRODUCTION

In establishing the fictitious crack model (FCM) for analyzing the mode-I type of cracks in concrete, Hillerborg et al. (1976) had also indicated the possibility of applying the concept to model other types of fracture, such as the shear fracture of the mode-II or mode-III type. Much effort has since been made to extend the fictitious crack model to mixed-mode fracture because most practical fracture problems in concrete are mixed-mode, involving modes I and II. In modeling the shear transfer mechanism in the fracture process zone (FPZ), numerical studies in this category often rely on interface elements: The stiffnesses of interface connections in the normal and tangential directions are assumed to be functions of the crack surface deformation (Bocca et al., 1991; Gerstle and Xie, 1992; Reich et al., 1993; Cervenka, 1994; Valente, 1995; Shah et al., 1995; Xie and Gerstle, 1995; Cervenka et al., 1998; Cendon et al., 2000; Galvez et al., 2002). However, this approach to introducing shear to the crack surface is inexplicit and approximate. An accurate specification of tangential tractions based on a given shear transfer law can hardly be achieved through regulating the stiffness of tangential connections that lack clear physical meaning. Since the shear transfer law defines the mode-II fracture energy, the lack of accuracy in implementing the law in numerical analysis could lead to erroneous results and misleading conclusions on the role and influence of the mode-II fracture parameters. Hence, a straightforward extension of the FCM is needed (Shi, 2004).

To clarify the shear transfer mechanism due to aggregate interlocking at crack surfaces in structural concrete, pioneering experimental investigations were carried out by many researchers, including Fenwick and Paulay (1968), Paulay and Loeber (1974), Fardis and Buyukozturk (1979), Bazant and Gambarova (1980), Bazant and Tsubaki (1980), and Reinhardt and Walraven (1982). Based on the obtained experimental evidence, Bazant and Gambarova conducted a thorough theoretical analysis on the stress-displacement relations on rough cracks and published a landmark paper in 1980 that offered insights and guidelines for numerical modeling of mixed-mode fracture that have been proved fundamental for understanding the nature of the phenomenon. Though their analysis was conducted on rough cracks in reinforced concrete, the same principles apply to cracks in plain concrete as well.

An outline of the analysis on the stress transfer mechanism at crack surfaces by Bazant and Gambarova is given following. Let δ_n and δ_t represent the relative normal and tangential displacements of the crack surfaces ($\delta_n \geq 0$). Here, δ_n is the crack opening displacement (COD), and δ_t is the crack sliding displacement (CSD). The cohesive forces contain two components: the normal stress σ_n and the shear stress σ_t, where subscripts n and t refer to the normal and tangential directions at the crack surface, respectively. The two force components are assumed to be functions of δ_n and δ_t, and are expressed as

$$\sigma_n = f_n(\delta_n, \delta_t) \tag{7.1}$$

$$\sigma_t = f_t(\delta_n, \delta_t) \tag{7.2}$$

According to Bazant and Gambarova, "For $\delta_n = 0$, there is full continuity in the material—that is, there is no crack. Thus, the states where $\delta_n = 0$ and $\delta_t \neq 0$ cannot be obtained." To ensure the finiteness of the fracture energy consumed or released by the crack as the displacement discontinuities increase from zero to δ_n and δ_t, the following work condition must be satisfied:

$$-\infty < W = \int_0^{\delta_n} \sigma_n d\delta_n + \int_0^{\delta_t} \sigma_t d\delta_t < \infty \tag{7.3}$$

Equation (7.3) restricts the admissible loading paths in the (δ_n, δ_t) plane as shown in Figure 7.1. Near the origin $\delta_n = \delta_t = 0$, a smooth and continuous path may be approximated by

$$\delta_t = c\delta_n^a \tag{7.4}$$

where c and a are constants. Bazant and Gambarova obtained empirical stress-displacement relations for Eqs. (7.1) and (7.2) by optimizing the fits of the available test data obtained by Pauley and Loeber (1974). Substituting these empirical relations and Eq. (7.4) into Eq. (7.3), they found that the fracture energy is bounded only if $a > 1$. This leads to

$$\frac{d\delta_t}{d\delta_n} = 0 \text{ for } \delta_n = 0 \tag{7.5}$$

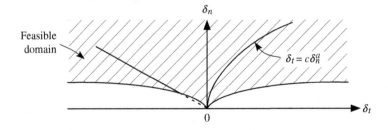

FIGURE 7.1 The feasible domain in the (δ_n, δ_t) plane for the normal and tangential stresses (after Bazant and Gambarova, 1980; courtesy of ASCE).

Based on this important finding, Bazant and Gambarova said the following:

> *Thus, the first displacement on the rough crack must be normal, and the slip can occur only after some finite opening has already been achieved. This condition must be carefully followed in numerical calculations. . . . Continuous cracks in concrete must propagate in such a direction that the displacement field near the crack tip be purely of mode-I (opening) type (i.e., the mode-II field cannot exist).*

Based on these analyses and a consideration of the actual existence of pure mode-I fracture in laboratory tests, the feasible domains for the normal and tangential stresses are illustrated in Figure 7.2 where δ_n^* denotes the critical COD for the onset of the CSD at crack surfaces. Possible tension-softening and shear-transfer relations of Eqs. (7.1) and (7.2) in the (δ_n, δ_t) plane are illustrated in Figure 7.3.

As shown, in general the normal and tangential components of the cohesive forces at the crack surface are functions of the displacements δ_n and δ_t. It should be noted that in Figure 7.3b, there is a delay in shear transfer in the initial stage of crack propagation. Obviously, this is associated with the late occurrence of the CSD that causes shear through aggregate interlocking. The delayed slip behind the crack tip can be understood by considering one fact: Since the CSD represents the slip between two crack surfaces, physically this term cannot precede the creation of the new crack surfaces by mode-I fracture at the crack tip. In this regard, mode-I fracture seems to be the only type of fracture that could take place independently because the creation of the new crack surfaces in the opening mode is not subject to such physical restrictions. For numerical analysis of mixed-mode fracture, it is crucial to introduce a delayed shear-transfer mechanism—in other words, to set a shear lag in a shear transfer law. As will be discussed later, in a numerical computation, removing the shear lag from a shear transfer function leads to invalid solutions.

In the following, modeling of the cohesive forces in the FPZ of a mixed-mode crack is addressed first. This is followed by reformulations of the FCM and the extended fictitious crack model (EFCM) for mixed-mode fracture, as well as a numerical formulation of the mode-II fracture energy. To verify the approach, single-notched shear beam tests by Arrea and

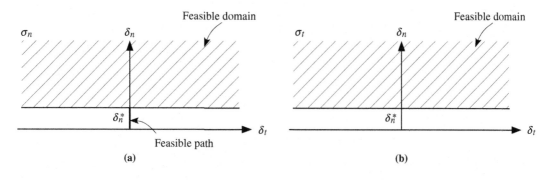

FIGURE 7.2 Feasible domains for σ_n and σ_t in the (δ_n, δ_t) plane: (a) feasible domain for σ_n and (b) feasible domain for σ_t.

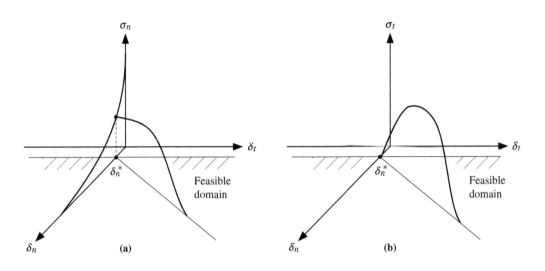

FIGURE 7.3 Concepts of tension-softening and shear-transfer relations in the (δ_n, δ_t) plane: (a) tension-softening relations in the (δ_n, δ_t) plane and (b) shear-transfer relations in the (δ_n, δ_t) plane.

Ingraffea (1982) are studied with five types of shear-transfer function, which reflect the variations of shear lag, shear strength, and the shape of the function. The obtained numerical results clearly reflect some of the cracking behaviors that were observed in the tests and convincingly demonstrate the validity of the FCM for modeling mixed-mode fracture of concrete. Based on these results, three more shear transfer functions are chosen for further studies, focusing on the mode-II fracture energy and its effect on the cracking behavior and the structural response. To apply the mixed-mode FCM and EFCM to engineering problems, the fracture tests on scale models of a gravity dam (Carpinteri et al., 1992), studied previously as mode-I fracture in Chapter 4, are remodeled as mixed-mode fracture, focusing on the influence of shear on the crack path in dams.

7.2 MODELING OF COHESIVE FORCES IN THE FPZ

In general, the constitutive relations for the stress and displacement in the FPZ of a mixed-mode crack are given by

$$\left\{ \begin{array}{c} d\sigma_n \\ d\sigma_t \end{array} \right\} = \left[\begin{array}{cc} D_{nn} & D_{nt} \\ D_{tn} & D_{tt} \end{array} \right] \left\{ \begin{array}{c} d\delta_n \\ d\delta_t \end{array} \right\} \tag{7.6}$$

in which [D] is the instantaneous moduli matrix that describes the tension-softening and shear-transfer relationships. It is known that the [D] matrix depends not only on the stress states but also on the loading path in the (δ_n, δ_t) plane. Based on the results of their tests, Paulay and Loeber (1974) observed that within certain ranges, the two displacement discontinuities have an approximately linear relation. Since slip can occur only after some finite opening of the crack has been achieved, the following relation is assumed:

$$\delta_t = \alpha\left(\delta_n - \delta_n^*\right) \tag{7.7}$$

Substituting Eq. (7.7) into Eq. (7.6) gives

$$\left\{ \begin{matrix} d\sigma_n \\ d\sigma_t \end{matrix} \right\} = \left\{ \begin{matrix} D_{nn} + \alpha D_{nt} \\ D_{tn} + \alpha D_{tt} \end{matrix} \right\} d\delta_n = \left\{ \begin{matrix} D'_{nn} \\ D'_{tn} \end{matrix} \right\} d\delta_n \tag{7.8}$$

where $D'_{nn} = D_{nn} + \alpha D_{nt}$, and $D'_{tn} = D_{tn} + \alpha D_{tt}$. According to Eq. (7.8), the tension-softening and shear-transfer laws can be defined by a single variable, δ_n.

Compared with the conceptually clear and theoretically well-established mode-I type of fracture in concrete, a unified approach to the mixed-mode fracture still does not exist, reflecting the complexity of the shear-transfer mechanism involved. Since there are no comprehensive mathematical models of the phenomenon, a functional approach based on Eq. (7.8) is adopted. First, it is assumed that the normal component of the cohesive forces is a unique function of the displacement discontinuity normal to the crack surface, effectively imposing the tension-softening law of concrete on the normal traction and the COD. As an example, the bilinear tension-softening relation in Figure 7.4 can be used for D'_{nn}, which is defined by the tensile strength of concrete f_t, the limit crack-opening displacement W_c^I, and the mode-I fracture energy G_F (Rokugo et al., 1989).

Second, for simplicity the tangential component of the cohesive forces, or shear, is also defined as a function of the COD, as shown in Figure 7.5. To be employed for D'_{tn}, the assumed bilinear and trilinear shear-COD relations possess the following characteristics. As the figure shows, there is a delay in shear transfer at the initial stage of crack propagation. As shear transfer begins at $W_{s1}(= \delta_n^*)$, the shear force builds up with increasing CODs and eventually attains its maximum value f_s at W_{s2}. Then, in the first two situations as shown in Figures 7.5a and b, it decreases with further opening of the crack (passing the concave point at W_{s3} in the case of the trilinear relation) and vanishes at W_c^{II}, marking the formation of an open crack. In another trilinear

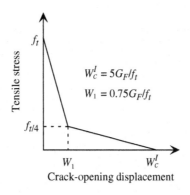

FIGURE 7.4 Bilinear tension-softening relation of concrete (Rokugo et al., 1989).

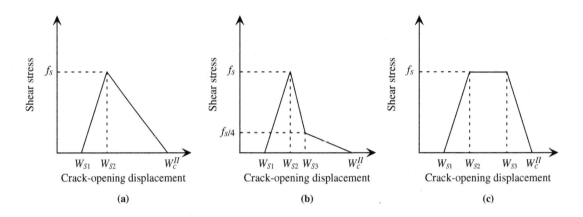

FIGURE 7.5 Proposed shear-COD relations with shear lag: (a) bilinear, (b) trilinear, and (c) trilinear with trapezoidal shape.

situation as shown in Figure 7.5c, the shear force does not decrease immediately after reaching the shear strength but retains that strength until the COD exceeds W_{s3}. After that, it decreases and vanishes at W_c^{II}.

This simple approach to shear transfer that is actually related with both the CSD and the COD has its theoretical basis in Eq. (7.8). In general, the limit crack-opening displacements for tension softening and shear transfer assume different values, W_c^I and W_c^{II}, because by definition shear transfer can still take place after the disappearance of the normal stress from the crack surface as the COD exceeds W_c^I. It should be pointed out that, unlike the mode-I case in Figure 7.4, where the enclosed area under the tension-softening curve represents the mode-I fracture energy G_F, the area under the shear-transfer curve in Figure 7.5 does not represent the mode-II fracture energy G_F^{II}, which can only be calculated through crack analysis based on the CSD. A numerical formulation of the mode-II fracture energy will be presented later.

7.3 REFORMULATION OF FCM AND EFCM FOR MIXED-MODE FRACTURE

In this section the FCM and the EFCM will be reformulated for mixed-mode fracture.

7.3.1 FCM for Mixed-Mode Fracture

Figure 7.6 shows a single crack of the mixed-mode type (I and II), propagating in the direction normal to the maximum principal tensile stress at the tip of the crack. In formulating crack equations, subscripts n and t represent, respectively, the normal and tangential components of force and displacement, and l stands for the limit value of a nodal force. Superscripts i and j denote

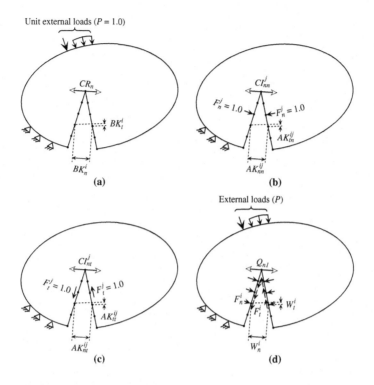

FIGURE 7.6 Concept at the crack-tip-controlled modeling of a single mixed-mode crack: (a) forces and displacements at the crack due to unit external loads; (b) forces and displacements at the crack due to a pair of unit normal forces; (c) forces and displacements at the crack due to a pair of unit shear forces; and (d) load condition for crack propagation.

the corresponding nodes at the crack. For the crack to propagate, the tensile force at the tip must reach the nodal force limit Q_{nl} (the tensile strength of concrete times the surface area apportioned to a nodal point), as shown in Figure 7.6d:

$$Q_{nl} = CR_n \cdot P + \sum_{i=1}^{M} CI_{nn}^i F_n^i + \sum_{i=1}^{M} CI_{nt}^i F_t^i \qquad (7.9)$$

where M is the number of nodes inside the fictitious crack. Here, CR_n, CI_{nn}^i, and CI_{nt}^i represent the nodal forces at the tip of the crack due to a unit external load, a pair of unit normal forces, and a pair of unit shear forces at the ith node of the crack, respectively. These coefficients are determined by linear-elastic FE computations based on the models from Figures 7.6a–c. Notice that P is the external load required to propagate the crack to the next nodal point.

In numerical analysis of mixed-mode fracture, the direction of the shear force must be carefully determined based on the crack surface deformation under the unit external load in

Figure 7.6a. Pairs of frictional forces are then applied against the surface sliding, as illustrated in Figure 7.6d. The COD and CSD at the ith node are obtained as

$$W_n^i = BK_n^i \cdot P + \sum_{j=1}^{M} AK_{nn}^{ij} F_n^j + \sum_{j=1}^{M} AK_{nt}^{ij} F_t^j \qquad (7.10)$$

$$W_t^i = BK_t^i \cdot P + \sum_{j=1}^{M} AK_{tn}^{ij} F_n^j + \sum_{j=1}^{M} AK_{tt}^{ij} F_t^j \qquad (7.11)$$

where $i = 1, \ldots, M$. Here, BK_n^i and BK_t^i are respectively the normal and tangential components of the compliance at the ith node due to the external load P. The influence coefficients AK_{nn}^{ij} and AK_{nt}^{ij} are the CODs at the ith node, and AK_{tn}^{ij} and AK_{tt}^{ij} are the CSDs, respectively, due to a pair of unit normal forces, and a pair of unit shear forces at the jth node of the crack. According to the reciprocity theorem, $AK_{nn}^{ij} = AK_{nn}^{ji}$, $AK_{nt}^{ij} = AK_{tn}^{ji}$, and $AK_{tt}^{ij} = AK_{tt}^{ji}$. FE models in Figure 7.6a–c are used to compute these coefficients.

Along the fictitious crack, the normal traction and the COD are assumed to follow the tension-softening law of concrete,

$$F_n^i = f_\sigma\left(W_n^i\right) \qquad (7.12)$$

and the shear and the COD are assumed to follow a shear-transfer law,

$$F_t^i = f_\tau\left(W_n^i\right) \qquad (7.13)$$

where $i = 1, \ldots, M$. Equations (7.9)–(7.13) establish the crack equations for propagating a single crack of the mixed-mode type, with the number of equations $(4M + 1)$ matching the number of unknowns $(4M + 1)$. By solving the crack equations, the unknown external load and the cohesive forces at the fictitious crack are determined, and the stress and displacement fields can be calculated under the revised boundary condition. The flowchart of the computational procedure is shown in Figure 7.7.

7.3.2 EFCM for Mixed-Mode Fracture

In the EFCM as proposed by Shi et al. (2001) for mode-I fracture, the approach to multiple-crack problems is to treat each crack individually as an active crack while restraining the growth of others and to establish relevant crack equations in terms of the unknown external loads, the CODs and the cohesive forces at the fictitious cracks. After solving the crack equations for each single-propagation scenario, a minimum load criterion is applied to identify the true active crack from among all the cracks involved, and stress analysis is carried out under the revised boundary condition. Apparently, this solution strategy applies to mixed-mode fracture, too.

Figure 7.8 illustrates two cracks of the mixed-mode (I and II) type—cracks A and B—propagating in the direction normal to the tensile force at the tip of the crack. In the following formulations, subscripts n and t represent the normal and tangential components of force and displacement, respectively, and l denotes the limit value of a nodal force. Superscripts a and b represent, respectively, cracks A and B, and i, j, and k denote the corresponding nodes at

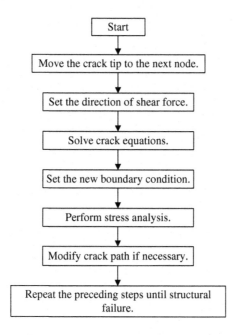

FIGURE 7.7 Solution procedure for the crack-tip-controlled modeling of a single mixed-mode crack.

designated cracks. For clarity, the cohesive forces and displacement discontinuities (COD and CSD) of the restrained cracks are marked by asterisks. When crack A is assumed to be active, the tensile force at its tip must reach the nodal force limit Q_{nl}^a (Figure 7.8f):

$$Q_{nl}^a = CR_n^a \cdot P^{(a)} + \sum_{i=1}^{N} CI_{nn}^{aai} F_n^{ai} + \sum_{i=1}^{N} CI_{nt}^{aai} F_t^{ai} + \sum_{j=1}^{M} CI_{nn}^{abj} F_n^{bj*} + \sum_{j=1}^{M} CI_{nt}^{abj} F_t^{bj*} \qquad (7.14)$$

where N and M are the number of nodes inside the two fictitious cracks, respectively. Here, CR_n^a, CI_{nn}^{aai}, CI_{nn}^{abj}, CI_{nt}^{aai}, and CI_{nt}^{abj} represent the nodal forces at the tip of crack A due to a unit external load, a pair of unit normal forces at the ith node of crack A and at the jth node of crack B, and a pair of unit shear forces at the ith node of crack A and at the jth node of crack B, respectively. These coefficients are obtained by linear-elastic FE calculations based on the models in Figures 7.8a–e. Notice that $P^{(a)}$ is the external load required to propagate crack A, while crack B remains inactive.

As stated earlier, the direction of the shear force at each crack must be determined from the tangential component of the crack surface deformation under the unit external load, as shown in Figure 7.8a. Pairs of shear forces are then applied against the surface sliding at each crack, as shown in Figures 7.8f and g. The CODs and CSDs along the two fictitious cracks are given by

$$W_n^{ai} = BK_n^{ai} \cdot P^{(a)} + \sum_{k=1}^{N} AK_{nn}^{aaik} F_n^{ak} + \sum_{k=1}^{N} AK_{nt}^{aaik} F_t^{ak} + \sum_{j=1}^{M} AK_{nn}^{abij} F_n^{bj*} + \sum_{j=1}^{M} AK_{nt}^{abij} F_t^{bj*} \qquad (7.15)$$

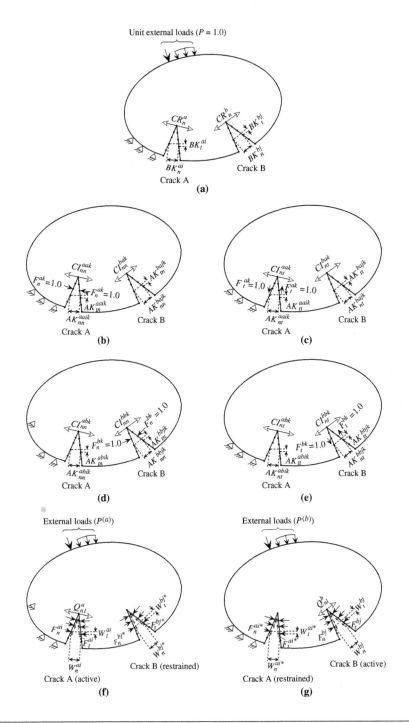

FIGURE 7.8 Concept of the crack-tip-controlled modeling of mixed-mode cracks: (a) forces and displacements at the cracks due to unit external loads; (b) forces and displacements at the cracks due to a pair of unit normal forces at crack A; (c) forces and displacements at the cracks due to a pair of unit shear forces at crack A; (d) forces and displacements at the cracks due to a pair of unit normal forces at crack B; (e) forces and displacements at the cracks due to a pair of unit shear forces at crack B; (f) load condition for the growth of crack A; and (g) load condition for the growth of crack B.

$$W_t^{ai} = BK_t^{ai} \cdot P^{(a)} + \sum_{k=1}^{N} AK_{tn}^{aaik} F_n^{ak} + \sum_{k=1}^{N} AK_{tt}^{aaik} F_t^{ak} + \sum_{j=1}^{M} AK_{tn}^{abij} F_n^{bj*} + \sum_{j=1}^{M} AK_{tt}^{abij} F_t^{bj*} \qquad (7.16)$$

$$W_n^{bj*} = BK_n^{bj} \cdot P^{(a)} + \sum_{i=1}^{N} AK_{nn}^{baji} F_n^{ai} + \sum_{i=1}^{N} AK_{nt}^{baji} F_t^{ai} + \sum_{k=1}^{M} AK_{nn}^{bbjk} F_n^{bk*} + \sum_{k=1}^{M} AK_{nt}^{bbjk} F_t^{bk*} \qquad (7.17)$$

$$W_t^{bj*} = BK_t^{bj} \cdot P^{(a)} + \sum_{i=1}^{N} AK_{tn}^{baji} F_n^{ai} + \sum_{i=1}^{N} AK_{tt}^{baji} F_t^{ai} + \sum_{k=1}^{M} AK_{tn}^{bbjk} F_n^{bk*} + \sum_{k=1}^{M} AK_{tt}^{bbjk} F_t^{bk*} \qquad (7.18)$$

where $i = 1, \ldots, N$ and $j = 1, \ldots, M$. Here, BK_t^{ai} and BK_n^{ai} at crack A, and BK_n^{bj} and BK_t^{bj} at crack B are the normal and tangential components of the compliances at nodes i and j, respectively, due to the external load $P^{(a)}$.

The influence coefficients AK_{nn}^{aaik}, AK_{nn}^{abij}, AK_{nt}^{aaik}, and AK_{nt}^{abij} are the CODs at the ith node of crack A due to a pair of unit normal forces at the kth node of crack A and at the jth node of crack B, and a pair of unit shear forces at the kth node of crack A and at the jth node of crack B, respectively. The influence coefficients AK_{tn}^{aaik}, AK_{tn}^{abij}, AK_{tt}^{aaik}, and AK_{tt}^{abij} are the CSDs at the ith node of crack A due to a pair of unit normal forces at the kth node of crack A and at the jth node of crack B, and a pair of unit shear forces at the kth node of crack A and at the jth node of crack B, respectively. Similarly, the influence coefficients AK_{nn}^{baji}, AK_{nn}^{bbjk}, AK_{nt}^{baji}, and AK_{nt}^{bbjk} represent the CODs, and AK_{tn}^{baji}, AK_{tn}^{bbjk}, AK_{tt}^{baji}, and AK_{tt}^{bbjk} stand for the CSDs both at the jth node of crack B, respectively, due to pairs of unit normal and shear forces at the corresponding locations. Based on the reciprocity theorem, $AK_{nn}^{aaik} = AK_{nn}^{aaki}$, $AK_{nn}^{bbjk} = AK_{nn}^{bbkj}$, $AK_{nn}^{abij} = AK_{nn}^{haji}$, $AK_{nt}^{aaik} = AK_{tn}^{aaki}$, $AK_{nt}^{bbjk} = AK_{tn}^{bbkj}$, $AK_{nt}^{abij} = AK_{nt}^{baji}$; and $AK_{tt}^{aaik} = AK_{tt}^{aaki}$, $AK_{tt}^{bbjk} = AK_{tt}^{bbkj}$, $AK_{tt}^{abij} = AK_{tt}^{baji}$. FE models to compute these coefficients can be found in Figures 7.8a–e.

Imposing the tension-softening law on the normal traction and the COD at each fictitious crack leads to

$$F_n^{ai} = f_\sigma\left(W_n^{ai}\right) \qquad (7.19)$$

$$F_n^{bj*} = f_\sigma\left(W_n^{bj*}\right) \qquad (7.20)$$

Similarly, a shear-transfer law is imposed on the shear and the COD:

$$F_t^{ai} = f_\tau\left(W_n^{ai}\right) \qquad (7.21)$$

$$F_t^{bj*} = f_\tau\left(W_n^{bj*}\right) \qquad (7.22)$$

Equations (7.14)–(7.22) establish the crack equations for mixed-mode fracture, and their solutions will specify the necessary conditions for crack A to propagate. With the number of equations $(4N + 4M + 1)$ matching the number of unknowns $(4N + 4M + 1)$, the problem can be solved uniquely, since these equations are linearly independent.

Alternatively, when crack B is assumed to be active (Figure 7.8g), the crack equations are readily obtained as

$$Q_{nl}^b = CR_n^b \cdot P^{(b)} + \sum_{i=1}^{N} CI_{nn}^{bai} F_n^{ai*} + \sum_{i=1}^{N} CI_{nt}^{bai} F_t^{ai*} + \sum_{j=1}^{M} CI_{nn}^{bbj} F_n^{bj} + \sum_{j=1}^{M} CI_{nt}^{bbj} F_t^{bj} \qquad (7.23)$$

$$W_n^{ai*} = BK_n^{ai} \cdot P^{(b)} + \sum_{k=1}^{N} AK_{nn}^{aaik} F_n^{ak*} + \sum_{k=1}^{N} AK_{nt}^{aaik} F_t^{ak*} + \sum_{j=1}^{M} AK_{nn}^{abij} F_n^{bj} + \sum_{j=1}^{M} AK_{nt}^{abij} F_t^{bj} \qquad (7.24)$$

$$W_t^{ai*} = BK_t^{ai} \cdot P^{(b)} + \sum_{k=1}^{N} AK_{tn}^{aaik} F_n^{ak*} + \sum_{k=1}^{N} AK_{tt}^{aaik} F_t^{ak*} + \sum_{j=1}^{M} AK_{tn}^{abij} F_n^{bj} + \sum_{j=1}^{M} AK_{tt}^{abij} F_t^{bj} \qquad (7.25)$$

$$W_n^{bj} = BK_n^{bj} \cdot P^{(b)} + \sum_{i=1}^{N} AK_{nn}^{baji} F_n^{ai*} + \sum_{i=1}^{N} AK_{nt}^{baji} F_t^{ai*} + \sum_{k=1}^{M} AK_{nn}^{bbjk} F_n^{bk} + \sum_{k=1}^{M} AK_{nt}^{bbjk} F_t^{bk} \qquad (7.26)$$

$$W_t^{bj} = BK_t^{bj} \cdot P^{(b)} + \sum_{i=1}^{N} AK_{tn}^{baji} F_n^{ai*} + \sum_{i=1}^{N} AK_{tt}^{baji} F_t^{ai*} + \sum_{k=1}^{M} AK_{tn}^{bbjk} F_n^{bk} + \sum_{k=1}^{M} AK_{tt}^{bbjk} F_t^{bk} \qquad (7.27)$$

$$F_n^{ai*} = f_\sigma \left(W_n^{ai*} \right) \qquad (7.28)$$

$$F_n^{bj} = f_\sigma \left(W_n^{bj} \right) \qquad (7.29)$$

$$F_t^{ai*} = f_\tau \left(W_n^{ai*} \right) \qquad (7.30)$$

$$F_t^{bj} = f_\tau \left(W_n^{bj} \right) \qquad (7.31)$$

where $P^{(b)}$ is the load required to propagate crack B, while crack A remains inactive.

By solving the two sets of crack equations—Eqs. (7.14) through (7.22) and Eqs. (7.23)–(7.31)—the true active crack is identified based on the minimum load criterion, $P = minimum$ $(P^{(a)}, P^{(b)})$, and stress analysis is carried out after adjusting the crack paths and modifying the force boundary condition accordingly. This process can be repeated until structural failure. Following the same solution approach, the preceding crack equations can be easily extended to include an arbitrary number of cracks. The flowchart of the computational procedure is shown in Figure 7.9.

As discussed in Chapter 4, the obtained numerical results must be checked to eliminate invalid solutions (due to irrelevant cracking modes assumed) upon solving the crack equations and obtaining the stress field. For details on evaluating the solution and predicting the crack path, refer to the previous discussion.

7.4 MODE-II FRACTURE ENERGY G_F^{II}

Suppose that the failure of a structural member is caused by a single crack of the mixed-mode type. Based on a step-by-step crack analysis, the total mode-II fracture energy is obtained by summing up the individual work performed by the shear force over the sliding distance at each node:

$$W_{total}^{II} = \sum_{i=1}^{N} W_i^{II} = \sum_{i=1}^{N} \sum_{j=1}^{M^i} F_t^i(T^j) \cdot [\delta_t^i(T^j) - \delta_t^i(T^{j-1})] \qquad (7.32)$$

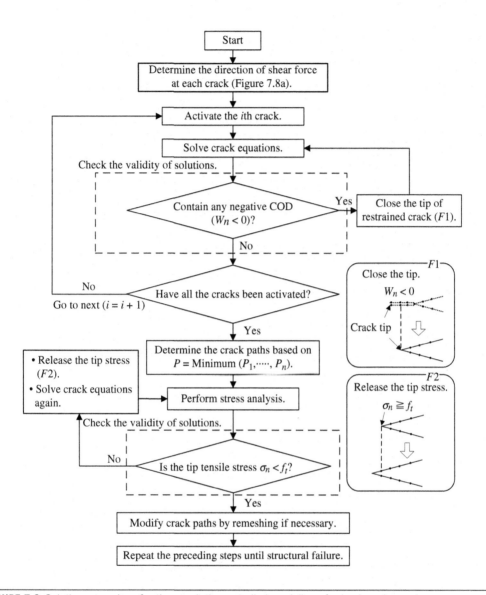

FIGURE 7.9 Solution procedure for the crack-tip-controlled modeling of mixed-mode cracks.

where N is the total number of nodes along the crack trajectory. M^i represents the total evolution-ary steps of the shear force at the ith node, and T^j is the jth evolutionary step. The mode-II fracture energy G_F^{II} is then obtained as

$$G_F^{II} = \frac{W_{total}^{II}}{A_{lig}} \tag{7.33}$$

where A_{lig} is the area of the ligament along the whole crack trajectory. Obviously, when multiple mixed-mode cracks are involved, the total fracture energy in Eq. (7.32) must be summed for each crack, and the fracture area A_{lig} should include all the crack trajectories to calculate the mode-II fracture energy G_F^{II}.

7.5 NUMERICAL STUDIES OF ARREA AND INGRAFFEA'S SINGLE-NOTCHED SHEAR BEAM

The single-notched shear beam test by Arrea and Ingraffea (1982) was truly a pioneering study of mixed-mode fracture that stimulated a wide range of discussions and extensive studies on shear fracture, shear strength, mode-II and mixed-mode fracture energy (Ingraffea and Panthaki, 1985; Watkins and Liu, 1985; Bazant and Pfeiffer, 1986; Davies et al., 1987; Davies, 1988; Barr and Derradj, 1990; Ballatore et al., 1990; di Prisco, et al., 2000). The geometry and loading arrangement of the selected test series can be found in the FE discretization of the beam specimen, as shown in Figure 7.10. In the experiment, the load was applied at point C of the steel beam AB and was controlled by a feedback mechanism with the crack mouth sliding displacement (CMSD) as a control parameter. As the notch is placed under nonsymmetric loading conditions, the crack propagating from the notch will exhibit opening as well as sliding on its crack surfaces. This unique fracture test has been studied numerically by many researchers using different approaches (Arrea and Ingraffea, 1982; Rots and de Borst, 1987; Rots, 1991; Gerstle and Xie, 1992; Nooru-Mohamed, 1992; Saleh and Aliabadi, 1995; Xie and Gerstle, 1995; Cendon et al.,

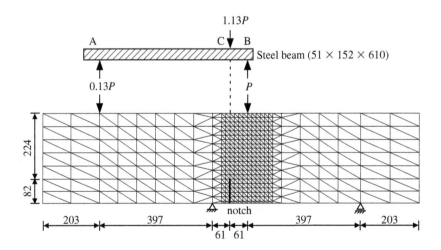

FIGURE 7.10 Finite element idealization (dimensions in mm) for single-notched shear beam tests by Arrea and Ingraffea (1982).

2000; Galvez et al., 2002; Ozbolt and Reinhardt, 2002). Due to limitations of the modeling methods employed, such as those of the interface element approach discussed previously, much remains to be clarified.

7.5.1 Parametric Studies with Five Shear-COD Relations

In the following studies, the bilinear tension-softening relation in Figure 7.4 is applied to the normal traction and the COD in the FPZ. Figure 7.11 illustrates five types of shear-COD relations to be used for the mixed-mode fracture analysis: Cases 1 to 3 are bilinear types, and Cases 4 and 5 are trilinear types. In all of the cases, the limit crack opening displacements for mode-I and mode-II stress transfer are assumed equal—that is, $W_c^{II} = W_c^I$. As seen, two types of shear strength are assumed.

For Cases 1, 3, 4, and 5, it is one-half of the tensile strength, $f_s = f_t/2$, and for Case 2, it is one-third, $f_s = f_t/3$. For Cases 1, 2, 4, and 5, the shear transfer begins at $W_{s1} = 0.1W_c^I$ and reaches its maximum at $W_{s2} = 0.2W_c^I$, while the concave point of Case 4 and the decreasing point of Case 5 are set at $W_{s3} = 0.3W_c^I$ and $W_{s3} = 0.4W_c^I$, respectively. For Case 3, a larger shear lag is assumed with $W_{s1} = 0.25W_c^I$, and the shear strength is reached at $W_{s2} = 0.5W_c^I$. Obviously, each of the five cases is chosen carefully to reveal how certain features in a shear-COD relation will affect the cracking behavior as well as the structural response. The material properties employed for the present study are shown in Table 7.1.

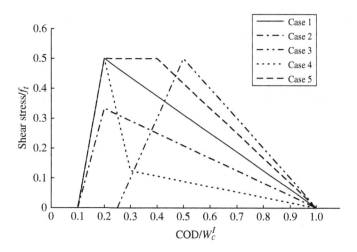

FIGURE 7.11 Five types of shear-COD relations: (a) Case 1 (bilinear: $f_s = f_t/2$, $W_{s1} = 0.1W_c^I$, $W_{s2} = 0.2W_c^I$, $W_c^{II} = W_c^I$); (b) Case 2 (bilinear: $f_s = f_t/3$, $W_{s1} = 0.1W_c^I$, $W_{s2} = 0.2W_c^I$, $W_c^{II} = W_c^I$); (c) Case 3 (bilinear: $f_s = f_t/2$, $W_{s1} = 0.25W_c^I$, $W_{s2} = 0.5W_c^I$, $W_c^{II} = W_c^I$); (d) Case 4 (trilinear: $f_s = f_t/2$, $W_{s1} = 0.1W_c^I$, $W_{s2} = 0.2W_c^I$, $W_{s3} = 0.3W_c^I$, $W_c^{II} = W_c^I$); and (e) Case 5 (trilinear: $f_s = f_t/2$, $W_{s1} = 0.1W_c^I$, $W_{s2} = 0.2W_c^I$, $W_{s3} = 0.4W_c^I$, $W_c^{II} = W_c^I$).

Table 7.1 Material Properties of Single-Notched Shear Beam

E (GPa)	ν	f_c (MPa)	f_t (MPa)	G_F (N/mm)
24.80	0.18	45.50	3.50	0.14

Figures 7.12 and 7.13 show the experimental envelope and the numerical predictions of the load versus CMSD curves and of the crack trajectories, respectively. For comparison, the numerical results obtained under the mode-I condition are also illustrated. As seen from the load-CMSD curves, the numerical analyses using the mode-I and mixed-mode conditions yield roughly the same peak loads (with variations of less than 3 percent) that agree reasonably well with the experimental results. The most obvious effect of the shear force on the global structural response appears in the postpeak regions. With the bilinear shear-COD relation of Case 1, the global stiffness increases greatly when compared with the observed structural response, and the material becomes relatively ductile. Apparently, lowering the shear strength in Case 1 will reduce this trend, as shown by the response curve of Case 2.

When the shear strength is reduced to zero, the mode-I condition is obtained, and the structural response becomes much more brittle. Upon imposing a larger shear lag in Case 3, a practically identical response curve of the mode-I condition is obtained up to a point beyond the peak load. As shear transfer takes place in the postpeak region, the rapid decrease of the load is interrupted by the local shear resistance at the crack surface, and the loading rebounds slightly as the structural

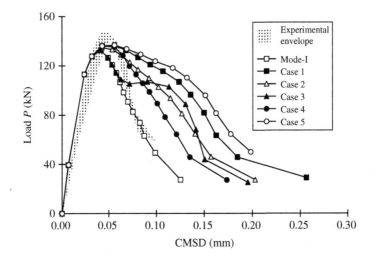

FIGURE 7.12 Experimental envelopes and numerical predictions of the load-CMSD curves under mode-1 and mixed-mode conditions with five types of shear-COD relations.

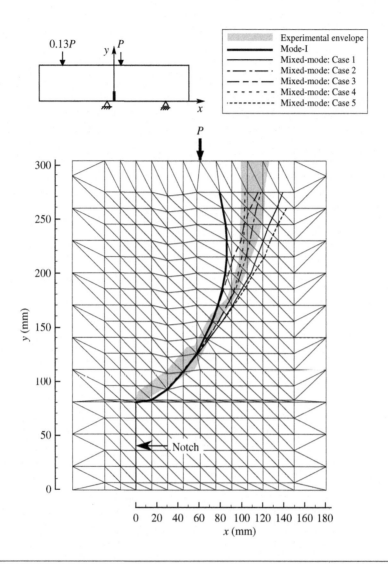

FIGURE 7.13 Experimental envelopes and numerical predictions of crack trajectories under mode-I and mixed-mode conditions with five types of shear-COD relations.

deformation progresses further before the final failure of the beam. With a trilinear shear-COD relation assumed in Case 4, where shear transfer takes place mainly in a narrow band of the COD, the response curve in the postpeak region moves closer to the experimental envelope. On the other hand, by introducing a trapezoidal type of transfer function in Case 5, the amount of shear transferred through the crack surfaces increases substantially, resulting in a response curve that is more ductile than that of Case 1.

Next, numerical results on the crack paths are examined. Figure 7.13 shows that the crack trajectories obtained with the mode-I condition and the mixed-mode conditions of Cases 1 and 5 are two opposite extremes, clearly deviating from the experimental envelope as the crack penetrates deeper into the beam. While the mode-I path suggests a splitting failure, by further increasing the shear strength in Cases 1 and 5 and delaying the decreasing point in the latter case, a shear type of fracture may be approached. As seen, the crack paths in Cases 2 and 4 fall into the band of experimental scatter. The cracking behavior of Case 3 is unique and highlights the effect of shear. As the decrease of the load in the absence of shear is temporarily halted by the rise of the shear force at the crack surface as just described, the crack path diverges immediately from the mode-I path and reenters the experimental envelope, as shown in Figure 7.13.

The preceding descriptions of the numerical results clearly demonstrate how a prescribed shear force in a mixed-mode fracture may affect the structural response and the crack path. As an additional check for the mixed-mode cases, the normalized normal and shear forces obtained at each node in the FPZ are shown in Figure 7.14. Notice that in each case a superimposition of the nodal forces obtained at all the crack-tip-controlled computational steps is presented, so Figure 7.14 exhibits the rise and fall of the nodal forces as the crack tip moves forward. Furthermore, the following relations at each node are shown in Figures 7.15 through 7.19 for Cases 1 to 5, respectively—that is, the relations of normal force versus COD, shear force versus COD, CSD versus COD, and shear force versus CSD.

The first two relations mirror the tension-softening and the shear-transfer laws (notice that the nodal force at the first node is small because the area apportioned to the initial node along a crack path is only one-half of the area apportioned to the other nodes, if an equal spacing between nodes is assumed). As shown in the CSD-COD relations, the CSD clearly lags behind the COD, causing the delay in shear transfer; as sliding takes place, these relations become more or less linear, hence justifying the linear approximation made in Eq. (7.7). Notice that the large CSDs at the first two nodes reflect the accumulation of the sliding displacements at other nodes along the crack trajectory.

Based on the obtained relations between the shear force and the CSD, the mode-II fracture energy is calculated step by step in Table 7.2. As seen, the values of G_F^{II} for Cases 1 to 5 are respectively 0.021, 0.014, 0.015, 0.010, and 0.024 N/mm. As expected, the largest G_F^{II} is obtained in Case 5, which is 17 percent of the mode-I fracture energy, G_F ($= 0.14$ N/mm). On the other hand, the smallest G_F^{II} is found in Case 4, which is merely 7 percent of G_F. Even though the mode-II fracture energy is small, the crack path with a mixed-mode condition is distinctively different from the mode-I path, as just shown and discussed.

It should be pointed out that in general, deleting the shear lag from the shear-transfer function in mixed-mode fracture would lead to a set of crack equations that have no valid solutions. In the present problem, by solving such equations, negative CODs would be obtained near the crack tip, suggesting that the single crack is closing under the external load, and thus the solution is invalid. As expected, the negative CODs are caused by the shear forces at the nodes close to the crack tip, due to the mixed-mode condition imposed on the displacement field near the crack tip. This finding is based on the results of numerical analyses and confirms the insightful assertion by Bazant and Gambarova (1980) that "continuous cracks in concrete must propagate in such a direction that the displacement field near the crack tip be purely of mode-I (opening) type (i.e., the mode-II field cannot exist)."

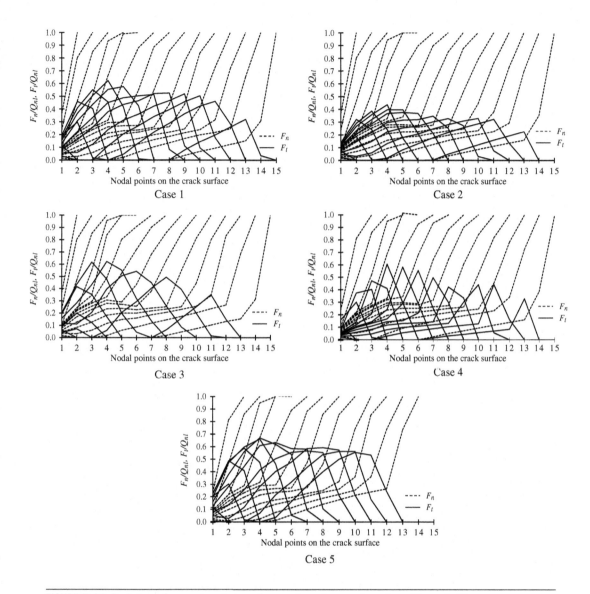

FIGURE 7.14 Evolution of normal and shear forces along the crack surface (Cases 1–5).

7.5.2 Parametric Studies on Mode-II Fracture Energy with Three Shear-COD Relations

Although the obtained mode-II fracture energy G_F^{II} varied significantly in Cases 1 to 5, the overall effect of G_F^{II} on the mixed-mode fracture was not discussed because other factors such as the shear lag and the type of transfer function may also have an influence on the numerical results. By modifying Case 2, whose crack path is deemed to be representative of the actual crack trajectories,

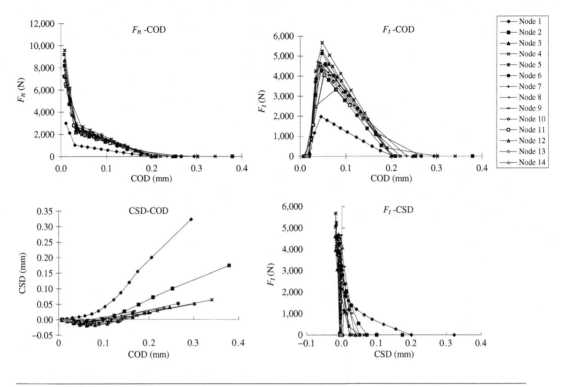

FIGURE 7.15 Relations of cohesive forces with COD and CSD, and relations of CSD and COD (Case 1).

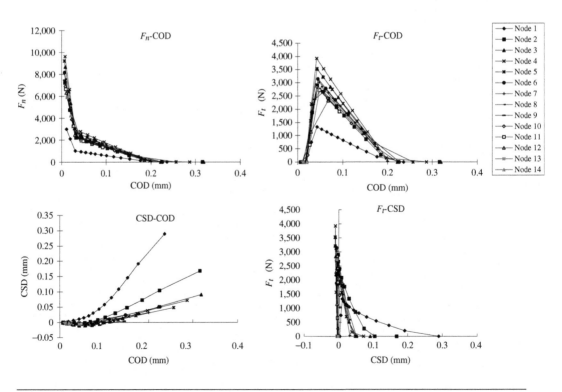

FIGURE 7.16 Relations of cohesive forces with COD and CSD, and relations of CSD and COD (Case 2).

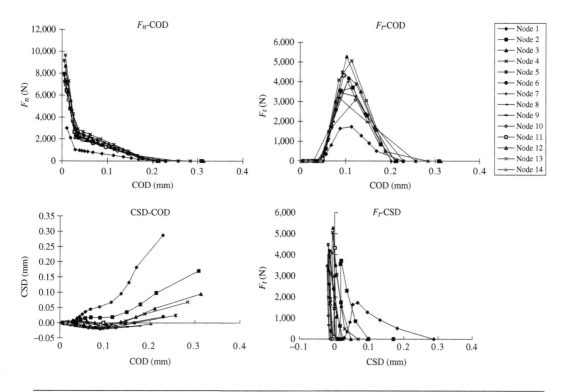

FIGURE 7.17 Relations of cohesive forces with COD and CSD, and relations of CSD and COD (Case 3).

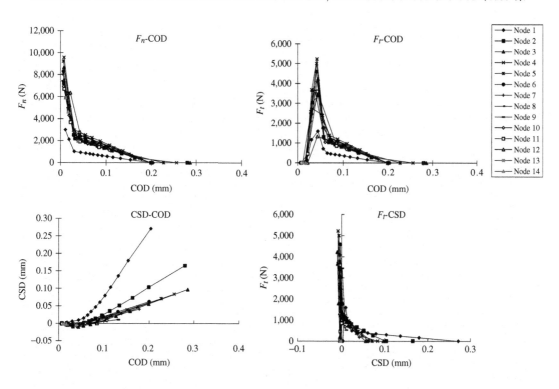

FIGURE 7.18 Relations of cohesive forces with COD and CSD, and relations of CSD and COD (Case 4).

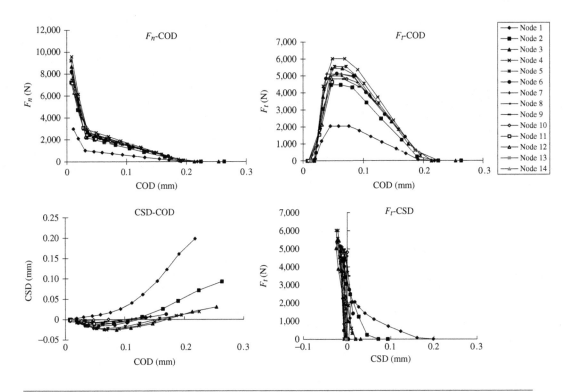

FIGURE 7.19 Relations of cohesive forces with COD and CSD, and relations of CSD and COD (Case 5).

three types of bilinear transfer function are derived and shown in Figure 7.20. With all of the other variables fixed, W_c^{II} is enlarged from the original value of W_c^I in Case 2 to $1.2W_c^I$, $1.5W_c^I$, and $2.0W_c^I$ in Cases 6 to 8, respectively, so as to increase G_F^{II}. Obviously, the following studies focus specifically on the mode-II fracture energy and its effect on the mixed-mode fracture.

Figure 7.21 presents the obtained load-CMSD relations, and Figure 7.22 illustrates the crack paths. The distributions of nodal forces along the crack surface are shown in Figure 7.23, which clearly exhibits shear transfer after the disappearance of the normal forces from the crack surface in Cases 6 to 8. This is in sharp contrast with Cases 1 to 5, in which the two force components disappear simultaneously from the crack surface due to the assumption of $W_c^I = W_c^{II}$, as can be verified in Figure 7.14.

The relations between nodal force and surface displacement and the CSD versus COD relations are shown in Figures 7.24 to 7.26. Based on the shear-CSD relations, the mode-II fracture energy is calculated for each case and shown in Table 7.2. As seen, the obtained G_F^{II} for Cases 6, 7, and 8 is 0.016, 0.019, and 0.024 N/mm, respectively. When compared with Case 2, in which $G_F^{II} = 0.014$ N/mm, the increase is respectively 14 percent, 36 percent, and 71 percent.

Considering the amount of the increase in G_F^{II}, the load-CMSD relations and the crack paths in Figures 7.21 and 7.22 are examined, in which the results of Case 2 are also illustrated. As shown, the increase of G_F^{II} in Cases 6 to 8 leads to response curves that show gentler decreasing tails and

Table 7.2 Computation of Mode-II Fracture Energy G_F^{II} [Eqs. (7.32) and (7.33)]

Item Case	W_1^{II}	W_2^{II}	W_3^{II}	W_4^{II}	W_5^{II}	W_6^{II}	W_7^{II}	W_8^{II}	W_9^{II}	W_{10}^{II}	W_{11}^{II}	W_{12}^{II}	W_{13}^{II}	W_{14}^{II}	W_{total}^{II} (N·mm) $\sum_{i=1}^{14} W_i^{II}$	A (mm)	B (mm)	$A \times B$ (mm²)	G_F^{II} (N/mm) $\dfrac{W_{total}^{II}}{A_{lig}}$
							W_i^{II} (N·mm)											$A_{lig} = A \times B$	
1	118.77	105.94	116.09	125.43	104.85	98.18	82.25	49.43	58.13	38.16	31.21	12.65	7.88	0.46	949.43	293.0	152.0	44536.0	0.021
2	108.59	83.68	70.24	74.03	62.56	58.06	38.96	45.51	21.07	17.20	9.69	5.77	8.16	0.00	603.51	284.7	152.0	43274.4	0.014
3	142.71	79.36	96.18	93.14	77.53	49.16	35.66	18.75	16.55	5.73	12.66	12.98	0.00	0.00	640.41	282.2	152.0	42894.4	0.015
4	61.13	58.00	58.25	57.44	48.93	41.44	30.03	33.59	13.75	14.65	7.92	5.38	11.81	0.00	442.31	282.5	152.0	42940.0	0.010
5	123.77	112.40	153.57	172.66	140.79	125.27	93.28	71.40	32.47	16.50	32.43	9.15	0.00	0.00	1083.68	296.8	152.0	45113.6	0.024
6	131.14	107.52	84.65	89.34	64.95	59.45	45.53	48.11	23.59	21.16	15.11	7.82	1.38	0.00	699.75	285.5	152.0	43396.0	0.016
7	174.67	141.57	102.20	98.46	74.56	73.06	48.60	47.32	30.05	22.39	10.92	5.56	13.83	0.47	843.64	288.2	152.0	43806.4	0.019
8	225.76	191.15	129.84	126.67	82.37	74.26	44.88	42.33	38.46	21.52	23.47	44.17	24.07	3.89	1072.85	290.4	152.0	44140.8	0.024

Note: A = ligament length; B = width of specimen

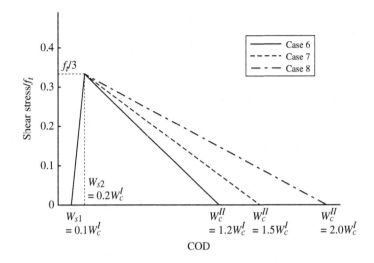

FIGURE 7.20 Three bilinear shear-COD relations in the study of mode-II fracture energy (Cases 6–8).

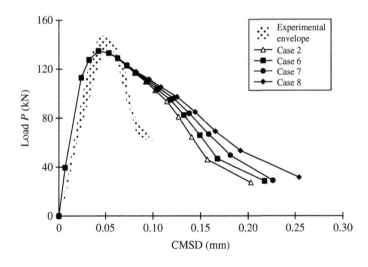

FIGURE 7.21 Experimental envelopes and numerical predictions of the load-CMSD curves under mixed-mode conditions (Cases 6–8).

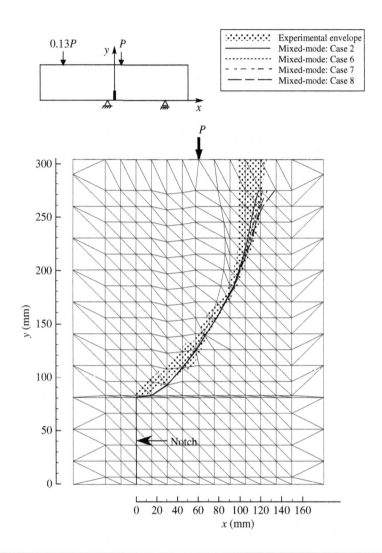

FIGURE 7.22 Experimental envelopes and numerical predictions of crack trajectories under mixed-mode conditions (Cases 6–8).

crack paths that deviate slightly from the crack path of Case 2 only in the last stage of crack propagation. Since the increase of G_F^{II} in these cases can be mainly attributed to the shear resistance in the region of large CODs ($> W_c^I$), its effect on the mixed-mode fracture seems to be quite limited. To gain a broader perspective on the issue, Case 1 and Case 2 are now compared also in terms of the mode-II fracture energy.

According to Table 7.2, the increase of G_F^{II} in Case 1 is 50 percent when compared to Case 2, which is due to the higher shear strength assumed in Case 1. Even though this increase is less than

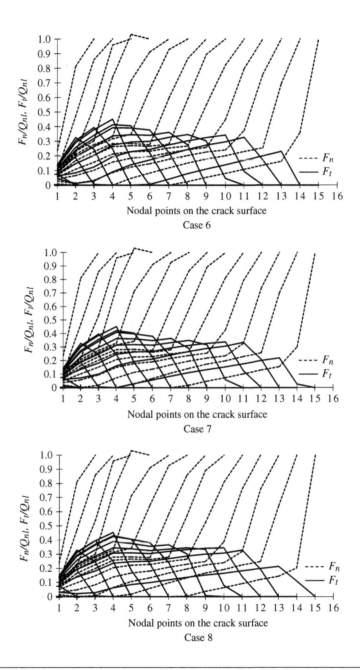

FIGURE 7.23 Evolution of normal and shear forces along the crack surface (Cases 6–8).

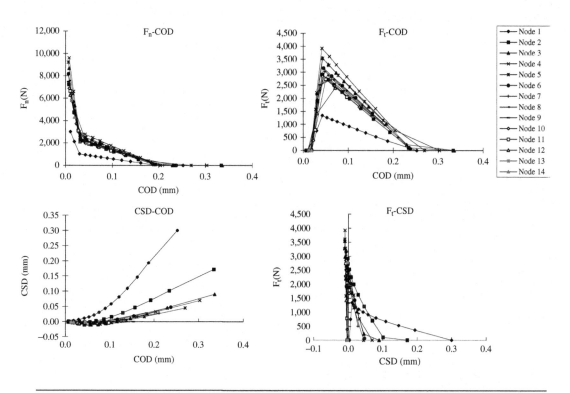

FIGURE 7.24 Relations of cohesive forces with COD and CSD, and relations of CSD and COD (Case 6).

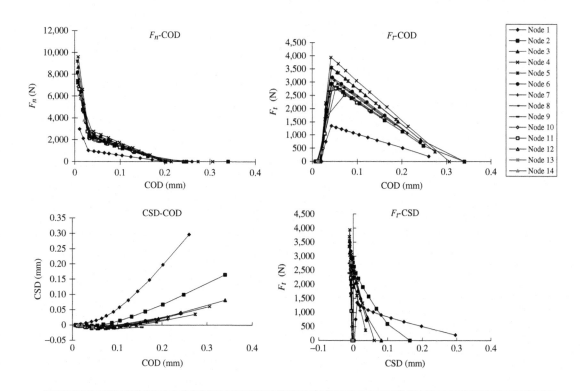

FIGURE 7.25 Relations of cohesive forces with COD and CSD, and relations of CSD and COD (Case 7).

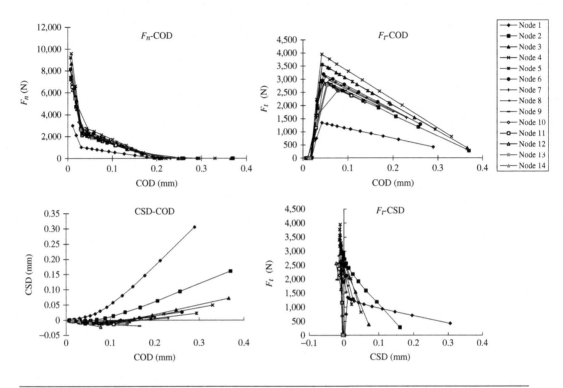

FIGURE 7.26 Relations of cohesive forces with COD and CSD, and relations of CSD and COD (Case 8).

the increase of 71 percent in Case 8, as shown in Figures 7.12 and 7.13, the cracking behavior and the structural response in Case 1 react strongly in the whole postpeak regions, which contrasts sharply with the limited response of Case 8 just described. This fact shows that the mode-II fracture energy spent in the region of small CODs ($< W_c^I$) has a much stronger influence on mixed-mode fracture than it does in the region of large CODs.

7.6 NUMERICAL STUDIES OF A SCALE-MODEL TEST OF A GRAVITY DAM

As clearly revealed by the preceding study on the fracture tests of single-notched shear beams, introducing shear on the crack surface may have a profound influence on the subsequent crack trajectory. In view of the sheer size and enormous gravity load of a dam structure, mixed-mode fracture must be carefully studied in the crack analysis of concrete dams: Neglecting shear from the crack surface may lead to an inaccurate prediction on crack path that may deviate from the true path significantly. The fracture test on scale-models of a concrete gravity dam under equivalent hydraulic loads by Carpinteri et al. (1992), which was discussed in Chapter 4 as mode-I fracture, is thus chosen as a mixed-mode fracture problem to be studied. The geometry and the loading arrangement of the test are shown in Figure 4.21.

To conduct the test, two 1:40 scale models of the dam were prepared, and for each specimen a horizontal notch was introduced into the upstream face located at a quarter of the dam height. With W representing the dam thickness at the height of a notch, the notch size varied from $0.1W$ in the first specimen to $0.2W$ in the second specimen. In total, three tests were carried out. Test 1 was unsuccessful due to an unstable failure along the base of the model, while an effort was made to simulate the self-weight condition. The repaired specimen was then fixed to the test platform and loaded to failure as Test 2. Under the same test condition as in Test 2, a fracture test was carried out on the second specimen as Test 3.

This well-known test was studied in Chapter 4 to analyze multiple cracking in dam structures using three FE models. Model I was the FE idealization of the original test, with two sizes of $0.1W$ and $0.2W$ assigned to the single notch. In Model II, two more notches were added above the single notch in Model I. In Model III, a different type of dam structure was assumed, which contained four notches and was subjected to a single load applied near the top. For brevity, only two of the three models are restudied in the following with mixed-mode conditions: Model I with the notch size of 0.2W (corresponding to Test 3) and Model III with two types of notch size combinations, as shown in Figure 7.27. In numerical studies, the bilinear tension-softening relation of Figure 7.4 is applied to the normal traction and the COD.

Two types of bilinear shear-COD relations are employed for the Model I analysis (Case 1: $f_s = 0.1f_t$, $W_{s1} = 0.1W_c^I$, $W_{s2} = 0.2W_c^I$, $W_c^{II} = W_c^I$; Case 2: $f_s = 0.2f_t$, $W_{s1} = 0.1W_c^I$, $W_{s2} = 0.2W_c^I$, $W_c^{II} = W_c^I$), and one type of bilinear shear-COD relation is assumed for the Model III analysis

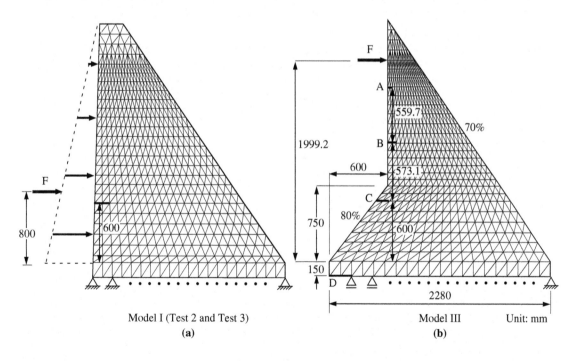

Model I (Test 2 and Test 3)   Model III   Unit: mm
(a)   (b)

FIGURE 7.27 Two numerical models of gravity dams with initial notches: (a) Model I and (b) Model III.

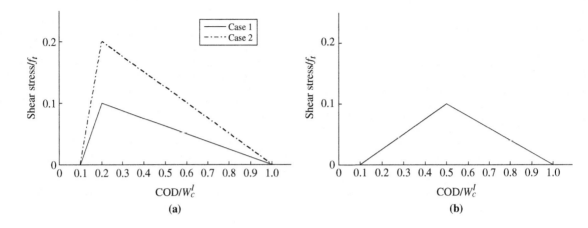

FIGURE 7.28 Bilinear shear-COD relations for (a) Model I and (b) Model III: (a) Model I (Case 1: $f_s = 0.1f_t$, $W_{s1} = 0.1W_c^I$, $W_{s2} = 0.2W_c^I$, $W_c^{II} = W_c^I$ and Case 2: $f_s = 0.2f_t$, $W_{s1} = 0.1W_c^I$, $W_{s2} = 0.2W_c^I$, $W_c^{II} = W_c^I$); and (b) Model III ($f_s = 0.1f_t$, $W_{s1} = 0.1W_c^I$, $W_{s2} = 0.5W_c^I$, $W_c^{II} = W_c^I$).

($f_s = 0.1f_t$, $W_{s1} = 0.1W_c^I$, $W_{s2} = 0.5W_c^I$, $W_c^{II} = W_c^I$), as shown in Figure 7.28. For the material properties of the test specimen, refer to Table 4.3.

Because the influence of shear force on the maximum load was found to be small and negligible in a preliminary study, the following discussion focuses on the crack path, which may be the foremost concern for the crack analysis of concrete dams. Figure 7.29 shows the obtained crack paths in Model I under the mode-I and the mixed-mode conditions, as well as the experimental observations. As seen, during the steady crack growth, the mode-I path follows a trajectory with a larger downturn angle (measured from the horizontal plane of the notch) than that of the observed crack path. Under the mixed-mode condition of Case 1 in which only a small shear strength $f_s = 0.1f_t$ is assumed, the crack propagates along the mode-I path initially; as shear transfer takes place, it diverges to a new path that lies in between the mode-I and the actual crack trajectories with a much smaller downturn angle. Increasing the shear strength in Case 2 to $f_s = 0.2f_t$, the crack propagates briefly along the mode-I path and then diverges to another path that merges with the observed crack trajectories.

These cracking behaviors are easy to comprehend. Under the mode-I condition, the discrepancy is caused by ignoring the frictional force or shear from the sliding surfaces of the crack, which is tantamount to omitting a local moment that acts against the surface deformation allowed under the mode-I condition. Here, the local moment is presumably formed by a pair of opposite shear forces acting on the two crack surfaces at the distance of a COD. Under the actions of these local moment forces in Cases 1 and 2, the crack paths eventually diverge from the mode-I path and move upward. As shown in the same figure, in the later stage of the unsteady crack propagation, totally different fracturing behaviors are predicted by the numerical analyses that do not agree

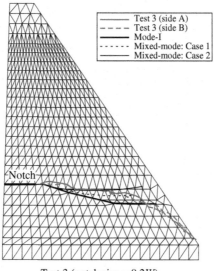

Test 3 (notch size = 0.2*W*)

FIGURE 7.29 Experimental results and numerical predictions of crack trajectories under mode-I and mixed-mode conditions with two types of shear-COD relations: Model I (Test 3; notch size = 0.2*W*) after (Shi, 2004; courtesy of ASCE).

with the experimental observations. This is believed to be caused by a transition of the loading conditions that may have occurred during the experiments and is not reflected in the numerical studies. For further details on this analysis, refer to section 4.7 of Chapter 4. The obtained normal and shear forces on the crack surface at all the computational steps in Cases 1 and 2 are shown in Figure 7.30.

Finally, the two cases in Model III that involve multiple cracking of the mixed-mode type are discussed. As shown in Figure 7.31, while the initial notches are basically assigned the same size of 0.1*W*, notch A of Case 1 and notch C of Case 2 are enlarged to 0.2*W*. Under the given load condition, cracks A and C become, respectively, the most active crack in each case. The curvilinear crack trajectories of these active cracks under the mode-I condition are also illustrated in Figure 7.31. Notice that the circled numbers along the crack path show the tip position of that particular crack at the designated step of the crack-tip-controlled computation. By introducing a mixed-mode condition with $f_s = 0.1f_t$, the obtained crack path in each case diverges from the respective mode-I path in the later stage of crack propagation, clearly restraining the sharp curving of the mode-I crack trajectory. The interactive growths of crack B in Case 1 and cracks A and B in Case 2 are also illustrated; these cracks remain as the mode-I type because their maximum CODs are less than $W_{s1} = 0.1W_c^l$, the assumed critical COD for shear transfer to take place. Figure 7.32 presents the evolution of nodal forces on the surface of the main crack in each case.

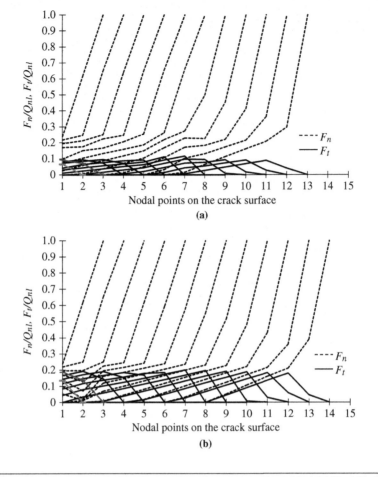

FIGURE 7.30 Evolution of normal and shear forces along the crack surface: Model I (Test 3): (a) Case 1 and (b) Case 2.

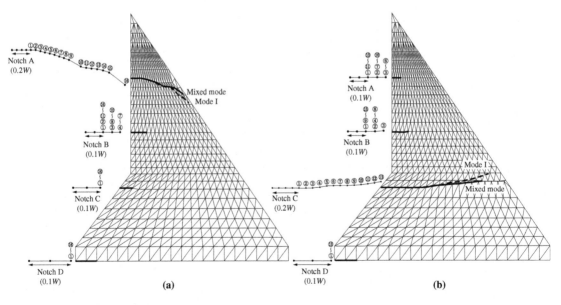

FIGURE 7.31 Numerical predictions of crack trajectories under mode-I and mixed-mode conditions: Model III (a) Case 1 and (b) Case 2 after (Shi, 2004; courtesy of ASCE).

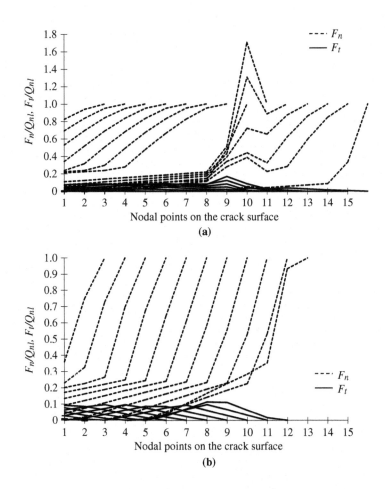

FIGURE 7.32 Evolution of normal and shear forces along the crack surface: Model III (a) Case 1 and (b) Case 2.

REFERENCES

Arrea, M., and Ingraffea, A. R. (1982). "Mixed-mode crack propagation in mortar and concrete." *Report No. 81-13*. Department of Structural Engineering., Cornell University, Ithaca, NY.

Ballatore, E., Carpinteri, A., Ferrara, G., and Melchiorri, G. (1990). "Mixed mode fracture energy of concrete." *Engineering Fracture Mechanics*, 35(1/2/3), 145–157.

Barr, B., and Derradj, M. (1990). "Numerical study of a shear (mode II) type test specimen geometry." *Engineering Fracture Mechanics*, 35(1/2/3), 171–180.

Bazant, Z.P. and Gambarova, P. (1980). "Rough cracks in reinforced concrete." *Journal of the Structural Division*, 106(4), 819–842.

Bazant, Z.P., and Pfeiffer, P. A. (1986). "Shear fracture tests of concrete." *Materials and Structures*, RILEM, 19, 111–121.

Bazant, Z.P., and Tsubaki, T. (1980). "Slip-dilatancy model for cracked reinforced concrete." *Journal of the Structural Division*, 106, ST9, 1947–1968.

Bocca, P., Carpinteri, A., and Valente, S. (1991). "Mixed mode fracture of concrete." *Int. J. Solids Structures*, 27(9), 1139–1153.

Carpinteri, A., Valente, S., Ferrara, G., and Imperato, L. (1992). "Experimental and numerical fracture modeling of a gravity dam." *Fracture Mechanics of Concrete Structures*, Z.P. Bazant, ed., pp. 351–360, Elsevier Applied Science.

Cendon, D. A., Galvez, J. C., Elices, M., and Planas, J. (2000). "Modeling the fracture of concrete under mixed loading." *International Journal of Fracture*, 103, 293–310.

Cervenka, J. (1994). *Discrete Crack Modeling in Concrete Structures*. PhD Thesis, University of Colorado, Boulder.

Cervenka, J., Chandra Kishen, J.M., and Saouma, V. E. (1998). "Mixed mode fracture of cementitious bimaterial interfaces; Part II: Numerical simulation." *Engineering Fracture Mechanics*, 60(1), 95–107.

Davies, J. (1988). "Numerical study of punch-through shear specimen in mode II testing for cementitious materials." *Int. J. Cement Composites and Lightweight Concrete*, 10(1), 3–14.

Davies, J., Yim, C. W. A., and Morgan, T. G. (1987). "Determination of fracture parameters of a punch-through shear specimen." *Int. J. Cement Composites and Lightweight Concrete*, 9(1), 33–41.

di Prisco, M., Ferrara, L., Meftah, F., Pamin, J., de Borst, R., Mazars, J., and Reynouard, J.M. (2000). "Mixed mode fracture in plain and reinforced concrete: some results on benchmark tests." *International Journal of Fracture*, 103, 127–148.

Fardis, M. N., and Buyukozturk, O. (1979). "Shear transfer model for reinforced concrete." *Journal of the Engineering Mechanics Division*, 105, EM2, 255–275.

Fenwick, R. C., and Paulay, T. (1968). "Mechanisms of shear resistance of concrete beams." *Journal of the Structural Division*, 94, ST10, 2325–2350.

Galvez, J. C., Cendon, D. A., and Planas, J. (2002). "Influence of shear parameters on mixed-mode fracture of concrete." *International Journal of Fracture*, 118, 163–189.

Gerstle, W. H., and Xie, M. (1992). "FEM modeling of fictitious crack propagation in concrete." *J. Engineering Mechanics*, 118(2), 416–434.

Hillerborg, A., Modeer, M., and Peterson, P. E. (1976). "Analysis of crack formation and crack growth in concrete by means of fracture mechanics and finite elements." *Cement and Concrete Research*, 6(6), 773–782.

Ingraffea, A. R. and Panthaki, M. J. (1985). "Analysis of shear fracture tests of concrete beams." *Proceedings US-Japan Seminar on Finite Element Analysis of Reinforced Concrete Structures*, pp. 71–91, Tokyo, May 21–24.

Nooru-Mohamed, M. B. (1992). *Mixed-Mode Fracture of Concrete: An Experimental Approach.* PhD Thesis, Delft University of Technology, Delft, The Netherlands.

Ozbolt, J., and Reinhardt, H. W. (2002). "Numerical study of mixed-mode fracture in concrete." *International Journal of Fracture*, 118, 145–161.

Paulay, T., and Loeber, P. J. (1974). "Shear transfer by aggregate interlock." *Shear in Reinforced Concrete, Special Publication* 42, 1, 1–15, ACI, Detroit.

Reich, R., Plizari, G., Cervenka, J., and Saouma, V. (1993). "Implementation and validation of a nonlinear fracture model in 2D/3D finite element code." *Numerical Models in Fracture Mechanics*, F. H. Wittmann ed., pp. 265–287, Balkema.

Reinhardt, H. W., and Walraven, J. C. (1982). "Cracks in concrete subject to shear." *Journal of the Structural Division*, 108, ST1, 207–224.

Rokugo, K., Iwasa, M., Suzuki, T., and Koyanagi, W. (1989). "Testing methods to determine tensile strain softening curve and fracture energy of concrete." *Fracture Toughness and Fracture Energy-Test Method for Concrete and Rock*, H. Mihashi, H. Takahashi, and F. H. Wittmann eds., pp. 153–163, Balkema.

Rots, J. G. (1991). "Smeared and discrete representations of localized fracture." *International Journal of Fracture*, 51, 45–59.

Rots, J. G., and de Borst, R. (1987). "Analysis of mixed-mode fracture in concrete." *Journal of Engineering Mechanics*, 113(11), 1739–1758.

Saleh, A., and Aliabadi, M. (1995). "Crack growth analysis in concrete using boundary element method." *Engineering Fracture Mechanics*, 51, 533–545.

Shah, S. P., Swartz, S. E., and Ouyang, C. (1995). "Nonlinear fracture mechanics for mode-I quasi-brittle fracture." *Fracture Mechanics of Concrete*, pp. 110–161, John Wiley & Sons,.

Shi, Z. (2004). "Numerical analysis of mixed-mode fracture in concrete using extended fictitious crack model." *J. Struct. Eng.*, 130(11), 1738–1747.

Shi, Z., Ohtsu, M., Suzuki, M., and Hibino, Y. (2001). "Numerical analysis of multiple cracks in concrete using the discrete approach." *J. Struct. Eng.*, 127(9), 1085–1091.

Shi, Z., Suzuki, M., and Nakano, M. (2003). "Numerical analysis of multiple discrete cracks in concrete dams using extended fictitious crack model." *J. Struct. Eng.*, 129(3), 324–336.

Valente, S. (1995). "On the cohesive crack model in mixed-mode conditions." *Fracture of Brittle Disordered Materials: Concrete Rock and Ceramics*, pp. 66–80, E & FN Spon.

Watkins, J., and Liu, K. L. W. (1985). "A finite element study of the short beam test specimen under mode II loading." *Int. J. Cement Composites and Lightweight Concrete*, 7(1), 39–47.

Xie, M., and Gerstle, W. H. (1995). "Energy-based cohesive crack propagation modeling." *J. Engineering Mechanics*, 121(12), 1349–1358.

Applications: Pseudoshell Model for Crack Analysis of Tunnel Linings

8.1 INTRODUCTION

In analyzing the structural behaviors and assessing the structural safety of aging tunnels such as railway tunnels, roadway tunnels, and waterway tunnels of hydraulic power facilities, available analytical approaches can be classified primarily into two categories. One is to model the time-dependent behavior of the surrounding geological materials with a simplified frame model for the tunnel lining (Kitagawa and Inagaki, 1993; Jiang et al., 1993; Zhang, 1994; Tasaka et al., 2000), in which hinges are introduced into the frame to model cracking in the lining concrete. The other is to model the lining with simplified boundary conditions, mostly with spring supports (Yin et al., 2001). In the latter approach, the smeared crack method is often used to model the nonlinear material behavior due to the occurrence of cracks in the lining. As is often observed during tunnel inspections, aging tunnels are often plagued with various kinds of cracking problems, such as those shown in Photo 8.1, where the crack-mouth-opening displacements (CMODs) of the two longitudinal cracks that were located respectively at the spring line and in the arch area reached several millimeters. In most of the cases, due to the lack of data measured on the cross-sectional deformation, these crack patterns and CMODs serve as important indices for evaluating the safety of these underground structures.

For the structural engineers in charge of the maintenance work of aging tunnels, it is desirable that the pressure loads acting on the lining can be determined from the CMODs, because the risk evaluations on these aging structures with clearly defined loading conditions will then become more objective than the practices that rely heavily on personal judgment. Apparently, neither of the preceding analytical approaches is readily available for this purpose, since cracks in these methods are not treated discretely. By virtue of the extended fictitious crack model (EFCM), a different approach has emerged that focuses on the cracking behavior and lining deformation. As an application of the EFCM, a pseudoshell model for crack analysis of the tunnel lining will be introduced in this chapter, which enables the CMODs of individual cracks to be calculated, corresponding deformations of the lining to be obtained, and pressure loads to be determined by a quasi loosening zone model.

From a structural point of view, a tunnel lining that contains through-thickness cracks like those in Photo 8.1 (cracks of this scale usually penetrate the whole depth of the wall) can hardly be considered as structurally stable without the interactive support from the surrounding

215

PHOTO 8.1 Longitudinal cracks observed in an aging waterway tunnel.

rock mass. To greatly simplify the problem, this shielding effect from the surrounding geological materials is replaced by a thin layer of pseudoshell, which is rigidly connected to the lining and assumes material properties equivalent to those of steel, as shown in Figures 8.1 and 8.2. As a result, the lining is stiffened.

This simplified approach is based on the beam theory that ensures the uniqueness of the solution on deformation when the beam is subjected to the same ratio of load to flexural rigidity. Therefore, by stiffening the lining in the pseudoshell model to increase its flexural rigidity, theoretically it is still possible to obtain the true lining deformation under certain conditions. To achieve this, crack analysis is indispensable. By applying relevant loads to the pseudoshell with the aim of reproducing the observed cracking behavior as shown in Figure 8.1, crack analysis is carried out based on the EFCM. When the tip of a crack reaches the pseudoshell, a plastic hinge is introduced into the shell, allowing a rigid-body rotation to take place at the crack surfaces while the crack analysis is continued using a crack-opening displacement (COD)-controlled algorithm. Hence, the cross-sectional deformation of the tunnel can be obtained at any specified CMODs. Next, a quasi loosening zone model is used to calculate the ground pressure. Based on the assumption that the ground deformation must equal the cross-sectional deformation of the tunnel, the external loads can be obtained by adjusting the extent of the loosening zone through iterative computations.

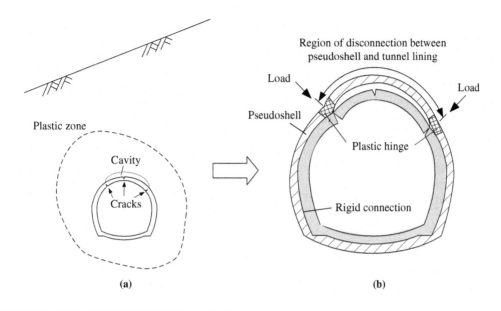

FIGURE 8.1 Concept of the pseudoshell model: (a) aging tunnel with longitudinal cracks and (b) pseudoshell model.

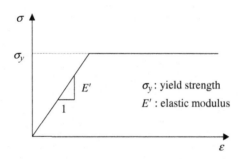

FIGURE 8.2 Elasto-plastic stress-strain relationship of the pseudoshell.

   In the following section, the beam theory, which forms the theoretical basis for the development of the pseudoshell model, is reviewed. As will be explained, to facilitate crack analysis in the pseudoshell model, the EFCM employs two modeling techniques. Before a crack penetrates the lining, crack analysis is carried out by the crack-tip-controlled modeling method, which was discussed in Chapter 4. As soon as the tip of a crack reaches the pseudoshell, a COD-controlled algorithm is required for the crack analysis, so in the following the EFCM is reformulated specifically for its application in the pseudoshell model. Parametric studies on the flexural rigidity of the pseudoshell are performed to verify the uniqueness of solutions on the lining deformation. To illustrate how the applied artificial load on the pseudoshell can affect the obtained cracking behavior, four types of dummy loads are used in these studies.

For further verification, a soil mechanics model is selected to analyze an aging waterway tunnel, which sustained multiple cracking in its lining and underwent a major renovation work more than 30 years after its construction. The same problem is then studied using the pseudoshell model to reproduce the recorded cracking patterns and the CMODs at the time of the renovation work. Comparisons between the two numerical results on the lining deformation at the time of renovation are made, and the results are found quite satisfactory. Based on the obtained lining deformation, the size of a quasi loosening zone, as well as the pressure loads, is calculated. Following this comparison study, four more sites in two aging waterway tunnels are chosen for further study, each representing a typical crack pattern frequently observed in various tunnels. Numerical results of the crack analysis in each case are compared with the maintenance records and discussed, and the pressure loads are calculated. Finally, a database is developed so that the ground pressure can be determined directly from the CMOD.

8.2 PSEUDOSHELL MODEL

In this section the modeling concept and numerical formulation of the pseudoshell model are presented, and parametric studies on the uniqueness of solutions on tunnel deformations are discussed.

8.2.1 Modeling Concept

As shown in Figure 8.3, for a beam element of arbitrary cross section, the governing equation for bending is derived as

$$\frac{d^2}{dx^2}\left(EI\frac{d^2v}{dx^2}\right) = q_z \tag{8.1}$$

where EI is the flexural rigidity, v is the displacement in the z-direction, and q_z is the distributed load acting perpendicularly to the beam axis. Assuming a constant flexural rigidity EI along the beam axis, Eq. (8.1) becomes

$$EI\frac{d^4v}{dx^4} = q_z$$

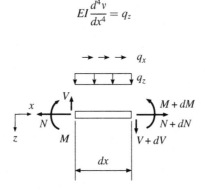

FIGURE 8.3 A beam element.

or

$$\frac{d^4v}{dx^4} = \frac{q_z}{EI} \tag{8.2}$$

Therefore, the beam deformation v becomes a function solely of the ratio of the external load q_z to the flexural rigidity EI. As long as this ratio is kept unchanged, the beam deformation is uniquely defined—that is, it is unaffected by the proportional variations of the load and the flexural rigidity of the beam. As shown following, this unique feature of the beam theory can be exploited to greatly simplify crack analysis problems in tunnels in which the interaction between the lining and the surrounding rock mass is too complex to be modeled realistically. Since the aim of the analysis is to obtain the pressure loads from the CMODs, it seems that a two-step solution can be used. Focusing on the bending action of the tunnel lining under the ground pressure, it is hoped that the cross-sectional deformation of the tunnel can be calculated using a simpler structural model with which detailed crack analysis can be carried out using the EFCM. Assuming that the tunnel deformation is thus obtained, the external loads can then be calculated using some simplified models in soil mechanics, which will be discussed later.

As shown in Figure 8.1, the pseudoshell model is thus proposed, which consists of the lining concrete and a thin layer of pseudoshell that assumes material properties equivalent to those of steel, as shown in Figure 8.2. Despite the fact that the flexural rigidity of the lining increases after rigidly connecting to an imaginary shell, the beam theory ensures the uniqueness of the solution on deformation with the same ratio of load to flexural rigidity. This offers an alternative means to calculate the true tunnel deformation using the pseudoshell model. As is illustrated in Figure 8.1, artificial loads are applied to the pseudoshell to propagate cracks in the lining, which should match the prescribed cracking behavior. Although the dummy loads represent the interactive forces between the lining and the rock mass, the magnitudes of these loads have no direct bearing on the actual loads. As a crack reaches the pseudoshell, yielding is enforced in a localized zone in the shell next to the crack, resembling the formation of a plastic hinge to allow rigid-body rotations at the crack surfaces.

With such a modeling concept, crack analysis can be carried out discretely, thus allowing any prescribed cracking behaviors to be reproduced in detail and the lining deformation to be calculated at the designated CMODs. Obviously, the theoretical justification for the obtained lining deformation lies in the assumption that the same ratio of load to flexural rigidity is obtained if the actual cracking behavior in a tunnel can be reasonably reproduced through crack analysis using the pseudoshell model. Of course, the validity of this assumption should be carefully verified in the following numerical studies.

8.2.2 Numerical Formulation

Once a crack penetrates through the lining, the resistance against bending substantially vanishes, and the crack propagation is replaced by a rigid-body rotation of the crack surfaces around the tip of the crack. To cope with this change of cracking mode, a COD-controlled modeling method is used to reformulate the EFCM. As the rotation of crack surfaces takes place, the ambiguity in determining the most active crack among multiple cracks vanishes, and the rotating crack is automatically treated as the leading crack. Figure 8.4 illustrates this situation with two cracks of the

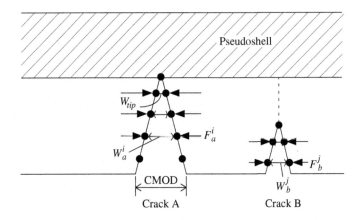

FIGURE 8.4 COD-controlled modeling of multiple cracks.

mode-I type, crack A and crack B, in which crack A has reached the pseudoshell, while crack B is still propagating in the lining concrete. In crack analysis, the rotation of the crack surfaces at crack A is achieved by monotonically increasing W_{tip}, which is the COD of the separated dual nodes next to the tip of the crack, as

$$W_{tip} = W_a^N = BK_a^N \cdot P + \sum_{k=1}^{N} AK_a^{Nk} F_a^k + \sum_{j=1}^{M} AK_{ab}^{Nj} F_b^j \tag{8.3}$$

where BK_a^N is the compliance at the Nth node of crack A due to the external load P. The influence coefficients AK_a^{Nk} and AK_{ab}^{Nj} are the CODs at the Nth node of crack A due to a pair of unit cohesive forces at the kth node of crack A and at the jth node of crack B, respectively.

Notice that W_{tip} in Eq. (8.3) is no longer an unknown but is a designated COD at the Nth node of crack A. Obviously, there is no more need to distinguish between the restrained and the active cracks in the present situation. The remaining CODs along the two fictitious cracks are given by

$$W_a^i = BK_a^i \cdot P + \sum_{k=1}^{N} AK_a^{ik} F_a^k + \sum_{j=1}^{M} AK_{ab}^{ij} F_b^j \tag{8.4}$$

$$W_b^j = BK_b^j \cdot P + \sum_{i=1}^{N} AK_{ba}^{ji} F_a^i + \sum_{k=1}^{M} AK_b^{jk} F_b^k \tag{8.5}$$

where $i = 1, 2, \ldots, N-1$, and $j = 1, 2, \ldots, M$. Finally, imposing the tension-softening law on the cohesive forces and the CODs leads to

$$F_a^i = f(W_a^i) \tag{8.6}$$

$$F_b^j = f(W_b^j) \tag{8.7}$$

where $i = 1, 2, \ldots, N$, and $j = 1, 2, \ldots, M$. The crack equations from (8.3) to (8.7) stipulate the conditions for the rotation of the crack surfaces at the tip of crack A and the growth of crack B in the lining. The problem can be uniquely solved because the number of equations $(2N + 2M)$, matches the number of unknowns, also $(2N + 2M)$. As stated before, numerical results obtained must be checked to eliminate invalid solutions, employing the steps described in Chapter 4. Obviously, the same procedure can be applied when crack B reaches the pseudoshell ahead of crack A, and it can be readily generalized to include any number of cracks.

It should be emphasized that, once W_{tip} exceeds the limit crack-opening displacement W_c, the leading crack then becomes a fully open crack; in other words, no more cohesive forces are transmitted through the crack surfaces. Thus, Eq. (8.3) is reduced to

$$W_{tip} = W_a^N = BK_a^N \cdot P + \sum_{j=1}^{M} AK_{ab}^{Nj} F_b^j \tag{8.8}$$

Notice that in actual computations involving several through-thickness cracks, the leading crack upon which W_{tip} is imposed may not be fixed to a specific crack. In a situation where the leading crack is found to be closing again because of the large opening of an adjacent crack, the role should be switched to another rotating crack.

The outline of the solution procedure is shown in Figure 8.5, which is composed of two separate analytical routines: the crack analysis routine and the stress analysis routine. For each given W_{tip}, the crack equations are solved first to determine the unknown external loads and the cohesive forces acting along each crack. After confirming the validities of the solutions, stress analysis is carried out. Notice that as a crack becomes a through-thickness crack, a plastic hinge is set in the pseudoshell. Following the stress analysis, the solution is checked again. The whole procedure is repeated until the designated CMODs are reached. It should be pointed out that in numerical analysis involving multiple cracks, there may be occasions when valid solutions do not exist for certain W_{tip}. Here, this problem is dealt with by simply skipping that particular W_{tip} and moving on to finding the solution for the next-larger W_{tip}.

8.2.3 Parametric Studies on Uniqueness of Solutions on Tunnel Deformation

The fracture test on the tunnel-lining specimen that was studied in Chapter 4 is chosen for parametric studies to verify the uniqueness of solutions on tunnel deformation, and its pseudoshell model is shown in Figure 8.6. Due to the assumption that a void exists above the ceiling, the pseudoshell is disconnected from the lining in that zone. For tunnels with caving problems, the frequently observed locations for crack propagation are illustrated on the half-model in Figure 8.6. As seen, these include crack A at the crown, growing from the outside of the lining, crack B in the arch area, and crack C at the spring line of the sidewall, growing from the inside of the lining.

The influence of the flexural rigidity of the stiffened lining on tunnel deformation is investigated by assuming three types of shell thickness $(t = {}^1/_5 H, {}^1/_3 H, {}^1/_2 H;\ H = $ lining thickness) with the same modulus of elasticity (E') and three types of moduli $(E', {}^1/_2 E', {}^1/_4 E')$ with the same shell thickness $(t = {}^1/_3 H)$. To investigate how a load condition can affect cracking behavior in the pseudoshell model, four types of dummy loads are considered: (I) distributed load along the shell; (II) two concentrated loads at B and C; (III) a concentrated load at B; and (IV) a concentrated load at C. For details, refer to Figure 8.6.

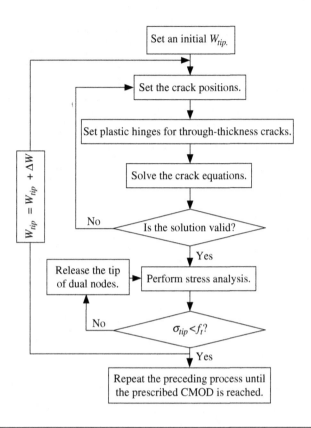

FIGURE 8.5 Solution procedure for crack analysis using the COD-controlled method.

In solving the crack equations, the bilinear tension-softening relation in Figure 4.9 is employed. Material properties of the lining specimen are summarized in Table 8.1, in which the assumed modulus of elasticity E' and Poisson's ratio v' for the pseudoshell are also listed. Notice that the yield strength is not assumed because a plastic hinge is automatically enforced in the pseudoshell as soon as the tip of a crack reaches the shell.

Figure 8.7 presents patterns of crack propagation obtained by using a pseudoshell of $t = \frac{1}{5}H$ under the four load conditions. As seen, with the distributed load (I) two cracks A and B are fully open at the crown and the edge of the void. A third crack C also appears at the spring line, but it closes early due to the active growth of crack B. The second (II) and third (III) types of loads result in similar cracking behaviors, except that no crack propagates from the spring line. With the fourth (IV) type of load, however, no crack extends from the edge of the void, and cracks develop at the crown and the spring line. These results suggest that crack patterns obtained by the pseudoshell model may not be very sensitive to the type of loads applied to the pseudoshell, which makes it easier to select proper dummy loads to produce a desired crack pattern.

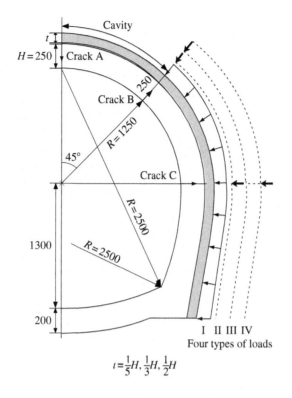

FIGURE 8.6 Numerical model for parametric studies ($t = \frac{1}{5}H, \frac{1}{3}H, \frac{1}{2}H$).

Table 8.1 Material Properties of Lining Concrete and Pseudoshell in Parametric Studies

Lining Concrete					Pseudoshell	
E (GPa)	ν	f_c (MPa)	f_t (MPa)	G_F (N/mm)	E' (GPa)	ν'
20.00	0.20	20.00	2.00	0.10	200.00	0.20

Relations between the tunnel deformation and the CMOD at the crown (crack A) and the edge of the void (crack B) are calculated under the distributed load conditions for the pseudoshells of different thicknesses in Figure 8.8, and for the pseudoshells of different moduli of elasticity in Figure 8.9. As clearly shown, these relations are uniquely defined and are unaffected by variations of the flexural rigidity of the stiffened lining, whether it is by changing the shell thickness or by changing its modulus of elasticity. This confirms the initial assumption of the pseudoshell model that the lining deformation under a prescribed crack condition can be uniquely obtained.

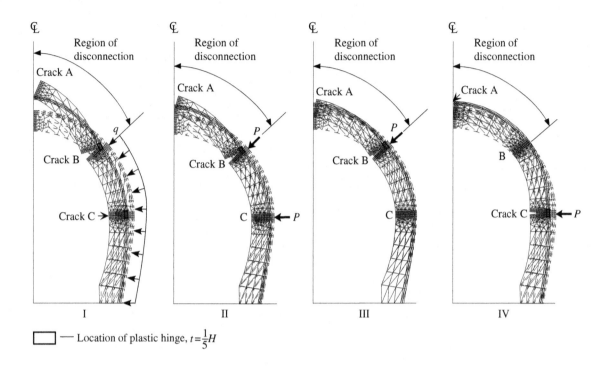

☐ — Location of plastic hinge, $t = \frac{1}{5}H$

FIGURE 8.7 Cracking behavior versus load type in parametric studies.

8.3 EVALUATION OF GROUND PRESSURE BASED ON THE QUASI LOOSENING ZONE MODEL

Ground pressure acting on tunnel linings can be defined as the external loads induced when that part of the ground in the vicinity of the tunnel endures large deformation due to tunneling as a result of the release and redistribution of the stress fields during and after tunnel excavation. Among the many theoretical approaches for estimating these pressure loads, which may involve different mechanisms for their occurrence, methods based on Terzaghi's theory on loosening zones have been widely used (Terzaghi, 1959). Here, a loosening zone may be understood as the region above the tunnel with clearly defined discontinuity from other areas after tunnel excavation, and the gravity loads in the region exert the so-called loosening ground pressure directly on the lining. It is known that the size and shape of the loosening zone vary according to the earth cover, geological structure and soil strength, internal friction angle, ϕ, and the shape and size of the tunnel. Since there is no general theory on the details of loosening zones, geological surveys and field measurements are often relied on for its estimation.

In the following, a simple model is defined for calculating the ground pressure, which assumes a simple form to facilitate iterative computations required by the solution procedures as described.

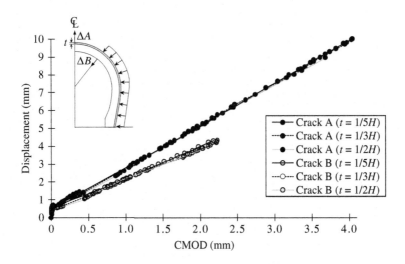

FIGURE 8.8 Cross-sectional deformation versus CMOD relations with the thickness of the pseudoshell as the variable.

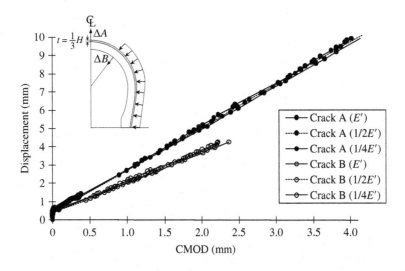

FIGURE 8.9 Cross-sectional deformation versus CMOD relations with the elastic modulus of the pseudoshell as the variable ($t = \frac{1}{3}H$).

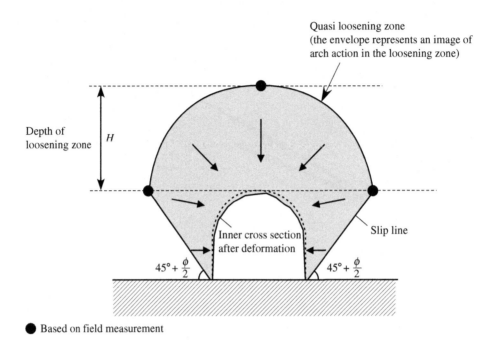

Depth of loosening zone H

Quasi loosening zone (the envelope represents an image of arch action in the loosening zone)

Inner cross section after deformation

Slip line

$45° + \dfrac{\phi}{2}$

$45° + \dfrac{\phi}{2}$

● Based on field measurement

FIGURE 8.10 Conceptual view of the quasi loosening zone model.

Named a quasi loosening zone model, it basically follows Terzaghi's concept of calculating the ground pressure from the depth of the loosening zone, even though other mechanisms might be involved. As shown in Figure 8.10, the model is defined by three points. The first two points are the locations where the two slip lines meet the horizontal line passing the crown (the slip surface is inclined at an angle of 45 degrees plus $\phi/2$ to the bottom of the tunnel). The third point represents the estimated depth of the loosening zone, usually based on field measurements. The three points are then connected by smooth curves to introduce the arch action in the ground.

If the ground and the lining are assumed to be in contact with each other (excluding areas where voids exist in between), the ground deformation must equal the cross-sectional deformation of the tunnel. Because the lining deformation can be calculated independently by crack analysis using the pseudoshell model, to determine the ground pressure, it is sufficient to modify the quasi loosening zone through iterative computations until the differences between the ground deformation and the lining deformation at key points of comparison (usually at inner cracks) can be ignored.

In numerical analysis, the quasi loosening zone is further simplified, as shown in Figure 8.11. Considering the zone an elastic body, the ground deformation is calculated under the vertical and lateral pressure loads. The task is to carefully adjust the depth of the loosening zone, the coefficient of lateral pressure, k, and other material properties until the ground deformation converges to the cross-sectional deformation of the tunnel. Solution procedures are illustrated in Figure 8.12.

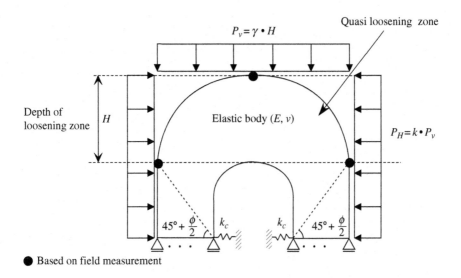

FIGURE 8.11 FE model of quasi loosening zone.

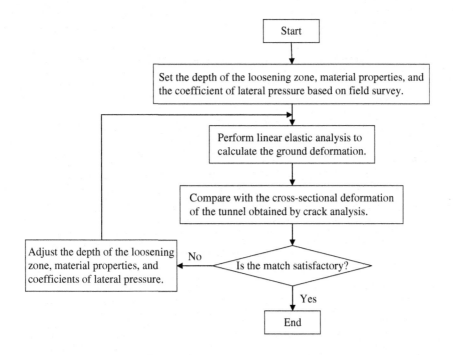

FIGURE 8.12 Solution procedures for calculating the pressure loads based on the quasi loosening zone model.

8.4 NUMERICAL ANALYSIS OF AN AGING WATERWAY TUNNEL (CASE A-1) COMPARED WITH A SOIL MECHANICS APPROACH

This section presents a comparison study on an aging waterway tunnel, which is analyzed first by a soil mechanics approach and then by the pseudoshell model.

8.4.1 Background

Numerical analyses of an aging waterway tunnel of a hydraulic power facility (power plant A) are carried out, and the numerical results are compared with those obtained previously using the Adachi-Oka model, which is an established constitutive model in geotechnical engineering (Adachi and Oka, 1992; Adachi et al., 1998). The model simulates the time-dependent behavior of geological materials that show strain-softening characteristics.

Figure 8.13 shows a cross section of the tunnel that was constructed in the early 1960s and was in service for over 30 years before major maintenance work was carried out. The tunnel passes under massive layers of unaltered sedimentary rocks of approximately 400 meters in depth. In its vicinity much more weakly consolidated sandstone and clay, which are extensively disturbed, exist. Based on the results of boring tests and PS logging, the loosening zone, which was formed during tunnel excavation, is estimated to reach a depth of 2 m. During excavation, a void was presumably formed above the ceiling area, and the records show that two longitudinal cracks nucleated in the arch areas on the lining surface shortly after the completion of the tunnel. Circumstantial evidence also points to the existence of another crack at the crown from the outer surface of the ceiling. At the time of the maintenance work, the CMODs of the two surface cracks had reached approximately 2 mm and 3 mm, respectively.

8.4.2 Numerical Analysis by Adachi-Oka Model

A detailed description of the Adachi-Oka model is beyond the scope of this book, so we present only a simple outline of the constitutive model. The shear strength of soft rock and over-consolidated clay consists of the strength due to cementation or bonding and the strength due to friction. With the gradual increase of shear strain, the former diminishes, while the latter grows. This process is manifested through strain softening. In Adachi-Oka's elasto-viscoplastic model, the stress history tensor is expressed by introducing a single exponential type of kernel function,

$$\sigma_{ij}^* = \frac{1}{\tau} \int_0^z \exp(-(z - z')/\tau)\sigma_{ij}(z')dz' \tag{8.9}$$

where τ is a material parameter that expresses the retardation of stress with respect to the time measure.

The incremental time measure is defined as

$$dz = g(\dot{\varepsilon}_{ij})dt \tag{8.10}$$

where g is an experimentally determined function of strain rate, and t is the time. Introducing a yield function and a nonassociated flow rule, a constitutive equation is derived to describe the

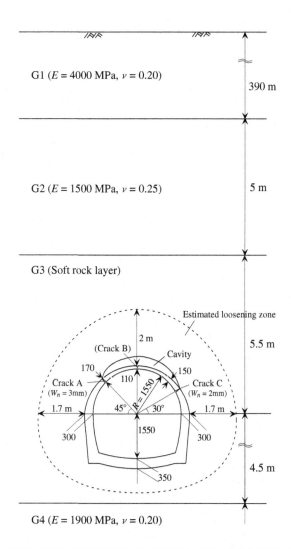

FIGURE 8.13 Cross section of an aging waterway tunnel (dimensions in mm) and geological conditions (Case A-1).

time-dependent stress-strain relation of geological materials. The constitutive model employs the following material parameters: E of the modulus of elasticity; v of Poisson's ratio; b and σ_{mb} of plastic potential parameters; G' and M_f^* of strain hardening-softening parameters; τ of a stress-history parameter; M_m of a parameter of overconsolidated boundary; and a and C of parameters of time dependency. These parameters are determined from conventional triaxial tests.

The main purpose of the original study using the Adachi-Oka model was to investigate the cross-sectional deformation of the tunnel lining and the time variation of earth pressure on the lining. The numerical analyses were carried out in the plane strain condition, and the tunnel

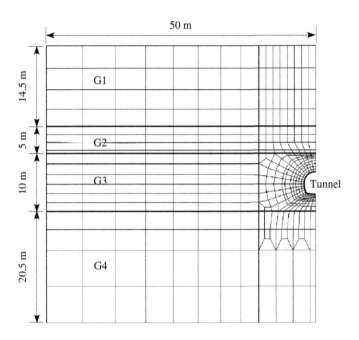

FIGURE 8.14 Finite element meshes in the elasto-viscoplastic modeling of Case A-1.

lining was modeled using beam elements. Figure 8.14 shows the finite element mesh, and Table 8.2 lists the parameters of the elasto-viscoplastic model. Table 8.3 summarizes the main parameters of the beam elements, where E is the modulus of elasticity, I is the moment of inertia, A is the cross-sectional area of the lining, and M_y is the moment of crack initiation. Notice that as the moment reaches M_y, a hinge is set to the beam at the location where the critical moment occurs.

The obtained numerical results were examined and verified by geotechnical surveys and in situ tests. Here, the time-dependent, cross-sectional deformation of the tunnel lining is shown in Figure 8.15. According to the numerical analyses, approximately two months after the completion of the tunnel, initial cracks appeared in the arch area and at the crown simultaneously. Approximately six years later, another crack occurred at the spring line. The final cross-sectional deformation after 32 years in service was estimated as 16.1 mm at the crown, 7.6 mm in the arch area, and 6.3 mm at the spring line.

Table 8.2 Material Parameters of Layer G3 in Elasto-Viscoplastic Modeling

E (MPa)	ν	b (MPa)	σ_{mb} (MPa)	G'	M_f^*	τ	M_m	a	C
300.00	0.25	0.87	18.00	45.40	1.15	90000	1.25	0.959	0.565

Table 8.3 Material Properties of the Beam Elements in Elasto-Viscoplastic Modeling

Member	E (GPa)	A (cm²)	I (cm⁴)	M_y (kN·m)
Lining	26.60	1500	28125.00	7.50
Invert	26.60	3500	357291.67	40.83

Note: Unit thickness = 1.0 m.

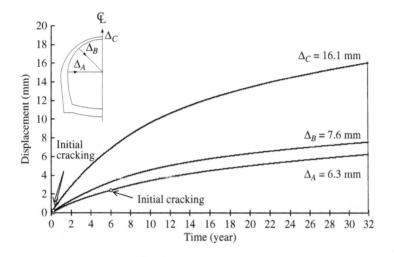

FIGURE 8.15 Time-dependent history of cross-sectional deformation obtained by the elasto-viscoplastic modeling of Case A-1.

8.4.3 Numerical Analysis by the Pseudoshell Model

Figure 8.16 presents the FE model of the cross section of the tunnel, excluding the invert. As seen, the void at the crown is not symmetric with respect to the central line but is inclining slightly toward the right. The lining thickness varies gradually from 300 mm at the spring line to just 110 mm at the crown. Here, the thickness of the pseudoshell is assumed to be 30 mm, just one-tenth of the lining thickness at the spring line. Notice that the lining is completely separated from the pseudoshell in the range of the void. At the bottom of the wall, hinge and spring supports are assumed in the vertical and horizontal directions, respectively. As illustrated, the numerical case contains three initial notches at the actual crack locations. Judging from the crack condition, two concentrated loads are applied to the pseudoshell, along the crack paths of the two surface cracks. The material properties used in numerical studies are summarized in Table 8.4, and the bilinear tension-softening relation in Figure 4.9 is employed to solve the crack equations.

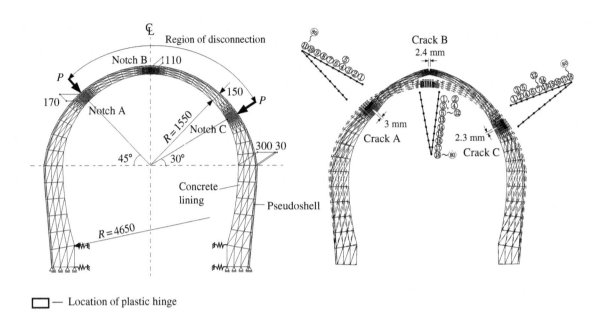

FIGURE 8.16 FE model and the results (dimensions in mm) of crack analysis (Case A-1).

Table 8.4 Material Properties of Lining Concrete and Pseudoshell in Case Studies

	Member	Lining Concrete					Pseudoshell	
Case	Item	E (GPa)	v	f_c (MPa)	f_t (MPa)	G_F (N/mm)	E' (GPa)	v
A-1		26.6	0.20	39.0	2.0	0.1	200.0	0.2
B-1		23.0	0.17	13.0	1.3	0.1	200.0	0.2
B-2		23.0	0.16	47.8	1.8	0.1	200.0	0.2
B-3		23.0	0.16	38.0	1.4	0.1	200.0	0.2
C-1		15.0	0.20	10.0	1.0	0.1	200.0	0.2

Also shown in Figure 8.16 are the results of crack analysis on crack propagation, with detailed information showing the tip position of each crack at the given computational step. As seen, crack A of the left arch is most active, progressing forward at every computational step, except for the 5th step when it becomes temporarily inactive as crack B becomes a one-step leading crack. As crack A becomes a through-thickness crack at the 11th step, subsequent computations are then carried out using the COD-controlled method. It is worth repeating that as a crack penetrates through the lining, a plastic hinge is introduced to the pseudoshell to allow the crack to fully open.

At the 18th step, the other two cracks penetrate through the lining simultaneously, and all the three cracks are open at this stage. Finally at the 80th step, the designated CMOD of 3 mm for crack A is reached, while the CMODs of crack B and crack C reach 2.4 mm and 2.3 mm, respectively. Notice that the obtained CMOD of crack C represents well its on-site measurement of approximately 2 mm.

Figure 8.17 presents the relations between the cross-sectional deformation and the CMOD. As shown, the obtained vertical displacement at the crown is 15.5 mm, and the arch deformations at A and C are 7.3 mm and 6.1 mm, respectively. Compared with the results obtained by the Adachi-Oka model, which predict a deformation of 16.1 mm at the crown and 7.6 mm in the arch area some 32 years after the completion of the tunnel, the agreement between the two approaches is indeed remarkable. These results convincingly prove the validity of the previous assumption that the lining deformation can be uniquely determined if the actual cracking behavior in a tunnel can be reproduced through crack analysis using the pseudoshell model.

For reference, the relations between the dummy load and the CMOD are shown in Figure 8.18. As seen, with all the cracks penetrating through the lining, the opening of crack surfaces is greatly accelerated, while the dummy load experiences a temporary decrease of more than 10 percent from its peak value due to the loss of the lining strength. The load increases again later, and its relation with the CMOD is approximately linear in the region of large CMODs.

In building a powerful constitutive model to describe the gross behaviors of geological materials, it is inevitable to employ a large number of material parameters to capture every important facet of the mechanisms that cause the overall behavior. Obviously, the accuracy of these material models depends very much on the accuracy of these material parameters. Most of them can only be determined through rigorous tests. The pseudoshell model, which is a unique structural model, presents a different approach. Exploiting the uniqueness of the solution on deformation in the beam theory, this approach focuses on the individual cracks in the lining, and the cross-sectional

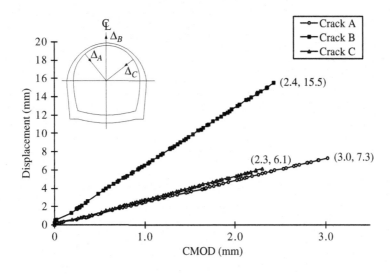

FIGURE 8.17 Cross-sectional deformation versus CMOD relations (Case A-1).

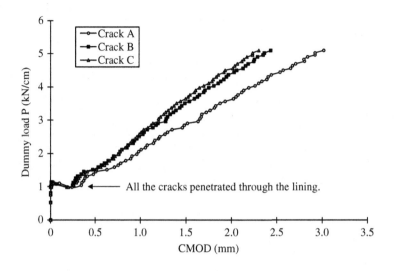

FIGURE 8.18 Dummy load versus CMOD relations (Case A-1).

deformation of the tunnel is obtained as a result of crack analysis, without considering the complicated details of geological materials that are of time-dependent nature.

8.4.4 **Evaluation of Ground Pressure**

Material properties of the rock mass in Case A-1 are shown in Table 8.5. To calculate the ground pressure by using the quasi loosening zone model, solution procedures illustrated in Figure 8.12 are followed. Based on the obtained lining deformations at cracks A and C, the depth of the loosening zone is adjusted first through iterative computations until a reasonable match is reached. Then the coefficient of lateral pressure and the loading range are also modified so the ground deformation in the sidewall sufficiently converges to the lining deformation there. Numerical results are shown in Figure 8.19, and the depth of the loosening zone is found to be 4 m, twice the initial estimation of 2 m. According to the analysis, the tunnel is under isotropic pressure loads of 0.096 MPa. The obtained lining and ground deformations are shown in Figure 8.20.

Table 8.5 Material Properties of Rock Mass in Case Studies

Case	γ (g/cm^3)	E (MPa)	ϕ (°)	ν
A-1	2.4	100	20	0.25
B-1	2.7	100	30	0.30
B-2	2.0	50	20	0.40
B-3	2.0	50	20	0.40
C-1	2.0	50	20	0.40

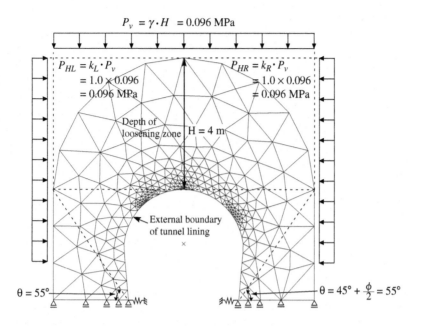

FIGURE 8.19 Obtained loosening zone and pressure loads (Case A-1).

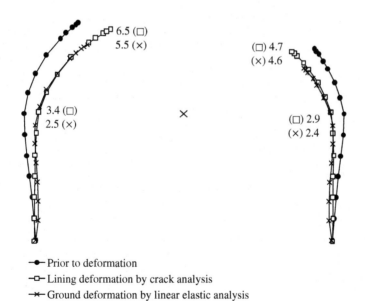

— Prior to deformation
— Lining deformation by crack analysis
— Ground deformation by linear elastic analysis

FIGURE 8.20 Obtained cross-sectional deformations by crack analysis (dimensions in mm) of the pseudoshell model and by linear elastic analysis of the quasi loosening zone model (Case A-1).

8.5 CASE STUDIES OF TWO AGING WATERWAY TUNNELS

Four more sites in two aging waterway tunnels are selected for further investigation, each representing a typical crack pattern frequently observed in tunnels with cracking problems:

1. An unsymmetric crack pattern, with cracks in the ceiling and sidewall (site B-1)
2. Two closely located cracks in sidewalls (site B-2)
3. A single crack in the ceiling (site B-3)
4. A symmetric crack pattern, with a single crack at each of the sidewalls (site C-1)

8.5.1 Power Plant B (Horseshoe Type): Site B-1

As shown in Figure 8.21, two longitudinal cracks were identified at site B-1, one in the ceiling area and another at the spring line of the right wall. The crack width was 3.7 mm in the ceiling and 5.4 mm at the spring line. The tunnel was located 180 m below the ground surface. According to the boring tests, the geological structure of the site was composed of claystone and sandstone alternations, and a cracking zone in the rock mass was found in the vicinity of the tunnel. Field measurements showed the depth of loosening zone to be 2 m, and a void was found behind the ceiling.

The FE model of Case B-1 is shown in Figure 8.22, which contains three crack paths. To propagate two inner cracks at locations A and C, a third crack must be introduced in the arch area, growing from the outside of the lining where a tension zone develops upon loading. Two concentrated loads are applied to the pseudoshell at A and C as dummy loads to propagate these cracks. Notice that the lining and shell are disconnected between A and C in numerical modeling. Material properties are summarized in Table 8.4 for all the cases.

The results of numerical analysis on crack propagation are shown in Figure 8.23. As seen, crack A is the leading crack, penetrating through the lining at the 10th step. Cracks B and C are less active at the initial stage of crack growth, and the two become through-thickness cracks at the 13th step simultaneously. The designated crack widths at A and C are obtained exactly at the 85th step by the COD-controlled computations, and the actual cracking mode at site B-1 is thus reproduced.

Other results are shown in Figures 8.24 and 8.25, which present the relations of the cross-sectional deformation and the CMOD and of the dummy load and the CMOD, respectively. As shown in Figure 8.24, despite the large crack width attained at crack C in the sidewall, the corresponding displacement of 2.5 mm at C is rather small compared with the displacements at the other locations. The displacement of 23.3 mm at B is the largest, approximately twice the displacement of 11.0 mm at A.

Table 8.5 presents the material properties of the surrounding rock mass for each case. Numerical results of the ground pressure analysis are shown in Figure 8.26, where the vertical pressure is 0.108 MPa, and the lateral pressure is 0.084 MPa from the right and 0.086 MPa from the left. The depth of loosening zone is obtained as 4 m, twice the field estimation of 2 m. Notice that the coefficient of lateral pressure on the left is slightly larger than that on the right, as a result of fine adjustment in iterative computations for the ground deformation to converge to the cross-sectional deformation of the lining in the region of contact. The obtained ground deformation is compared with the lining deformation in Figure 8.27.

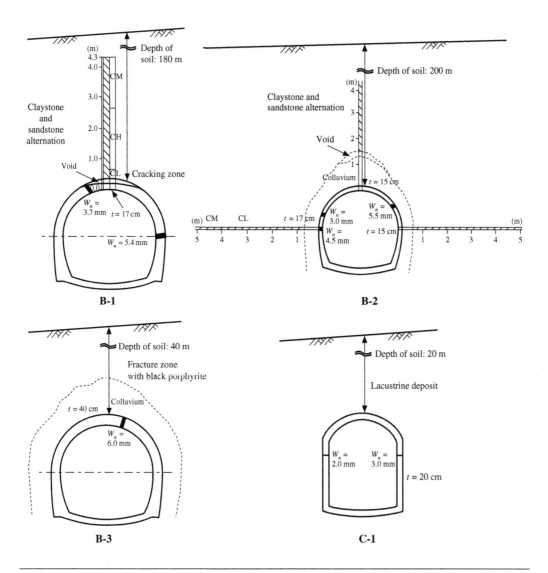

FIGURE 8.21 Cross sections of aging waterway tunnels and geological conditions in case studies.

8.5.2 **Power Plant B (Horseshoe Type): Site B-2**

As shown in Figure 8.21, two closely located cracks were found in the left wall, with the crack widths reaching 4.5 mm and 3.0 mm, respectively, and the third one was identified in the arch area on the right, with the CMOD reaching 5.5 mm. Based on the field investigation, the present case was limited to a small section of the tunnel, and for most of the cracked sections only one crack appeared in the same wall, along with a crack in the arch area on the right. According to

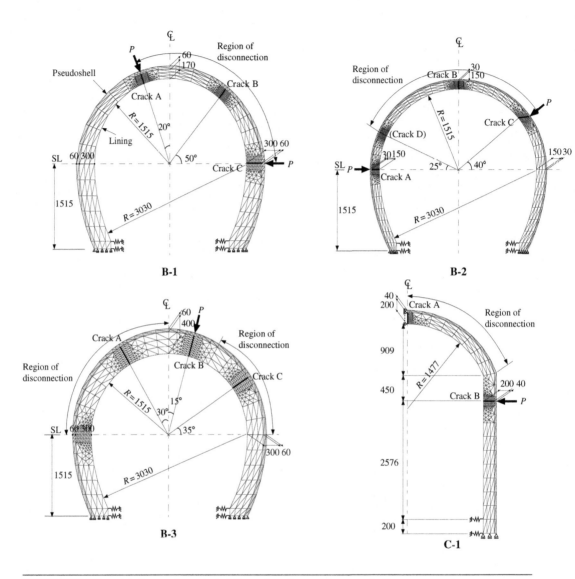

FIGURE 8.22 FE models for crack analysis (dimensions in mm) in case studies.

the boring tests, the geological structure of the site was claystone and sandstone alternations. The earth cover above the tunnel was 200 m. The estimated depth of loosening zone was 2 m by the field tests, and behind the ceiling, a lump of colluvium was found with voids.

In a preliminary FE analysis, it was found that to propagate two closely positioned cracks in the sidewall, the load type and loading position must be changed at some stages of the numerical computation. Without doing so, the opening of one crack would force the other to close, and vice versa. To simplify the situation, the most active crack at the spring line is considered, and the FE

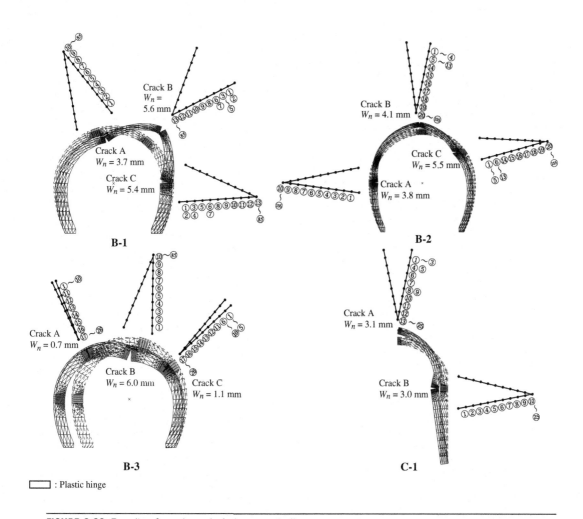

FIGURE 8.23 Results of crack analysis in case studies.

model is as shown in Figure 8.22. As seen, in addition to the crack paths set for the two inner cracks, another has to be introduced at the crown to propagate a crack from the outside. Two dummy loads are applied on the pseudoshell at A and C, and the region of disconnection between the lining and shell is from C to D.

The results of crack analysis are shown in Figure 8.23. As seen, crack A is the dominant crack and penetrates through the lining at the 10th step. Cracks B and C are almost nonpropagating at the initial stage, but they become active at the 14th step and penetrate through the lining at the 20th step simultaneously. The designated crack width of 5.5 mm at C is obtained at the 155th step, and the corresponding CMOD at A is 3.8 mm, slightly smaller than the target value of 4.5 mm. This discrepancy seems to be caused by the simplification of the actual cracking behavior in the sidewall. As shown in Figure 8.24, the final displacement is 11.3 mm at A,

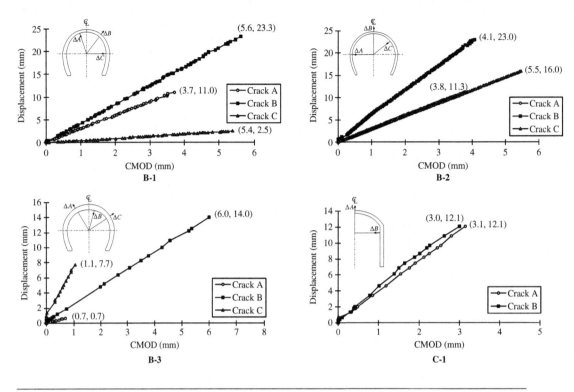

FIGURE 8.24 Cross-sectional deformation versus CMOD relations in case studies.

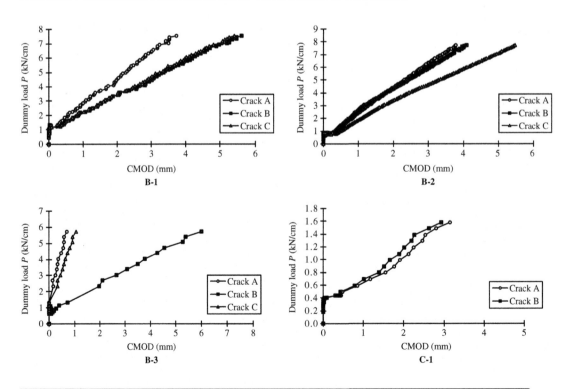

FIGURE 8.25 Dummy load versus CMOD relations in case studies.

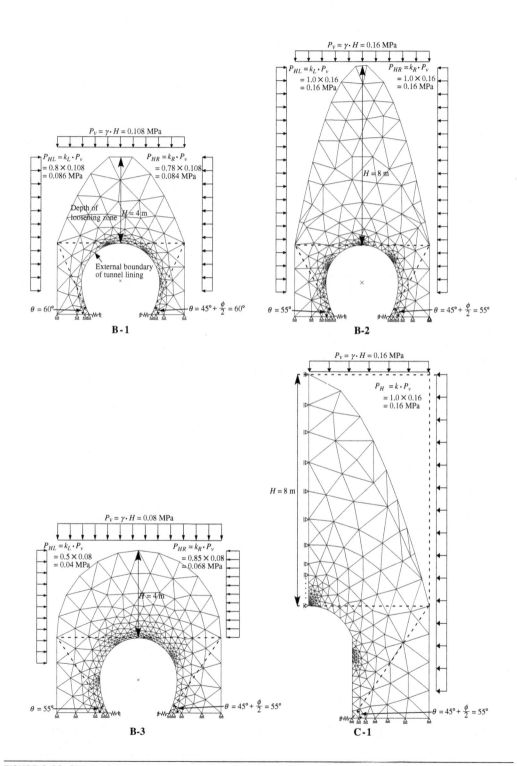

FIGURE 8.26 Obtained loosening zones and pressure loads in case studies.

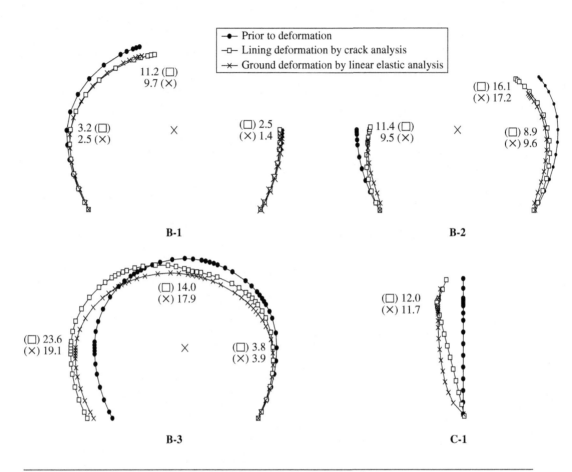

FIGURE 8.27 Obtained cross-sectional deformations by crack analysis (dimensions in mm) of the pseudoshell model and by linear elastic analysis of the quasi loosening zone model in case studies.

23.0 mm at B, and 16.0 mm at C. For the relations between the dummy load and the CMOD, refer to Figure 8.25.

Numerical results on the external loads are illustrated in Figure 8.26, and the obtained depth of loosening zone is 8 m, far exceeding the field estimation of 2 m. Based on the analysis, the tunnel is under an isotropic earth pressure of 0.16 MPa. The obtained cross-sectional deformations by the two methods are compared in Figure 8.27.

8.5.3 Power Plant B (Horseshoe Type): Site B-3

At site B-3, a longitudinal crack was found in the ceiling, and its crack width reached 6 mm, as shown in Figure 8.21. According to the records, this section was sandwiched by neighboring linings that had longitudinal cracks in the arch areas. Based on the boring tests, the site was located in a fracture zone with black porphyrite, and the tunnel was surrounded by thick layers of colluvium. The depth of the soil above the tunnel was 40 m, and no void was found behind the ceiling. The maintenance records did not offer any information on the depth of loosening zone for this site.

The FE model is shown in Figure 8.22, where three crack paths are prescribed, one for the inner crack in the ceiling and two for the outer cracks in the arch areas. The locations of the outer cracks are assumed to be the same as those of the arch cracks in the neighboring linings. As seen, a dummy load is applied to the shell at crack B, and the shell is disconnected from the lining around the two outer cracks to allow their propagations. The results of crack analysis are shown in Figure 8.23. As seen, crack B is the leading crack, and its crack width reaches 6.0 mm at the 26th step.

In numerical computations, to gain a large opening for crack B but with only limited openings of cracks A and C, the horizontal support at the bottom of the wall on the left must be set free. To do otherwise (i.e., to set free the horizontal support on the right) would lead to inappropriate cracking behavior and large tensile stresses on the surface of the wall. The obtained displacement is 0.7 mm at crack A, 14.0 mm at crack B, and 7.7 mm at crack C, as shown in Figure 8.24. The relations between the dummy load and the CMOD are given in Figure 8.25. Figure 8.26 presents the results of ground pressure analysis, which predict a vertical pressure of 0.08 MPa and a lateral pressure of 0.068 MPa on the right and 0.04 MPa on the left, and the depth of the loosening zone is obtained as 4 m. The resulting cross-sectional deformation is compared with the lining deformation in Figure 8.27.

8.5.4 **Power Plant C (Calash Type): Site C-1**

As shown in Figure 8.21, at site C-1 a longitudinal crack appeared in each sidewall, and the crack width was 2.0 mm on the left and 3.0 mm on the right. The geological structure of the site was mainly of claystone and sandstone alternations, identified as lacustrine deposit. The earth cover above the tunnel was 20 m, and no voids were found between the lining and the surrounding rock mass. Like the previous case, no field tests were carried out to estimate the depth of the loosening zone.

Figure 8.22 presents a half-FE-model for crack analysis, simplifying the problem based on a consideration of approximate symmetry. As seen, in addition to the crack path set in the sidewall, another path is assumed at the crown. To propagate this crack from the outside, the lining is disconnected from the shell in the ceiling area. A single load is applied to the pseudoshell at crack B. The results of crack analysis are shown in Figure 8.23. As expected, crack B in the sidewall is the leading crack, which becomes a through-thickness crack at the 10th step, and its crack width reaches the designated value of 3.0 mm at the 25th step. On the other hand, the propagation of crack A at the crown is hindered by some nonpropagating steps before it penetrates through the lining at the 13th step. The obtained lining deformations at cracks A and B are found to be the same as 12.1 mm, as shown in Figure 8.24. The relations between the dummy load and the CMOD are shown in Figure 8.25.

Figure 8.26 illustrates the obtained quasi loosening zone with a depth of 8 m, and the tunnel is found to be under isotropic pressure loads of 0.16 MPa. The obtained ground deformation is compared with the lining deformation in Figure 8.27. It should be pointed out that by assuming a smaller loosening zone and a larger coefficient of lateral pressure, one more solution on the ground pressure can be found in this case. In a situation like this, a solution is then determined based on the severity of the possible load conditions, unless the less severe condition can be proved as proper.

8.6 DEVELOPMENT OF DATABASE FOR EVALUATION OF GROUND PRESSURE BASED ON THE CMOD

Through the case studies (sites A1, B1-B3 and C-1) just discussed, it has been demonstrated that for tunnels with cracked linings, the pseudoshell model and the quasi loosening zone model can work together to provide an effective means for calculating the ground pressure based on the CMOD. To explore the potential of this approach, in the following a database is developed for selected aging waterway tunnels to determine the external loads based on the CMOD. Such information is valuable for developing an effective maintenance system for the safe operation of aging tunnels of any kind.

8.6.1 Selection of Influential Factors and Cases of Study

In this section influential factors for the development of the database are discussed, and cases for study are selected.

Type of Tunnel and Structural Dimensions

Two types of waterway tunnel are considered: the horseshoe type and the calash type, both shown in Figure 8.28. The structural dimensions of the two model tunnels are determined based on the standard design for waterway tunnels with an inner diameter of 3 m.

Crack Pattern

Among the many actual crack patterns observed in aging waterway tunnels, three typical crack patterns are chosen for each of the two model tunnels, as shown in Figure 8.29. These cracking modes are symmetric with respect to the central line of the tunnel, and the locations

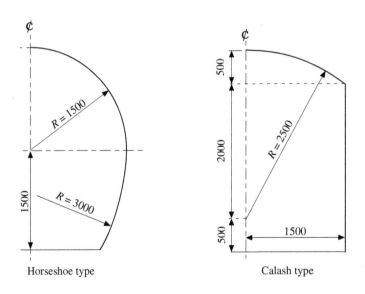

FIGURE 8.28 Geometric dimensions (in mm) of two model tunnels for database development. (In some countries the model on the left is called the "Oval tunnel" and the model on the right is the "Horseshoe tunnel.")

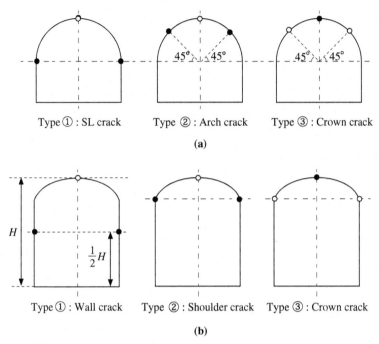

FIGURE 8.29 Idealized crack patterns in tunnel linings considered for database development: (a) horseshoe type and (b) calash type.

of cracks are the most frequently observed places that are prone to cracking in each type of structure. For the horseshoe type, these are the spring lines, the arch areas, and the crown, and for simplicity the corresponding cracks are termed the SL crack, the arch crack, and the crown crack. For the calash type, these are the two points at half of the tunnel height in the sidewalls, the shoulders of the sidewalls, and the crown, and for simplicity the corresponding cracks are termed the wall crack, the shoulder crack, and the crown crack. Notice that in Figure 8.29 the inner cracks are denoted by the black dots, and the outer cracks are represented by the white dots.

Thickness of Lining

Since the thickness of the lining is an important structural geometry for tunnels, three types of lining thickness are considered for database development: 20 cm, 30 cm, and 50 cm. These values represent the majority of the lining thicknesses in aging waterway tunnels that were measured by coring tests and recorded in numerous field investigations.

Classification of Geological Materials

The magnitude of ground pressure acting on the tunnel depends very much on the properties of the geological materials that surround the tunnel. Based on the previous case studies, the rock mass is classified into three grades for database development, and for each class typical values of the material properties are assumed in Table 8.6.

Depth of the Quasi Loosening Zone

According to the quasi loosening zone model, the pressure loads depend directly on the depth of the quasi loosening zone. To obtain the relations between ground pressure and cross-sectional deformation, various values for the depth of loosening zone are assumed in ground pressure analysis, and the range of change is from 1 m to 10 m.

Cases of Study

Based on the preceding classification of influential parameters that affect cracking behavior and ground pressure, numerical cases for crack analysis are listed in Table 8.7. As shown, these cases are the combinations of tunnel type, crack pattern, and thickness of the lining. The designated crack widths for calculating the cross-sectional deformation of the tunnel are 1 mm, 3 mm, and 5 mm, respectively. The numerical cases for ground pressure analysis are presented in Table 8.8,

Table 8.6 Assumed Material Properties of Rock Mass for Database Development

Classification of Rocks	γ (g/cm^3)	E (MPa)	v	ϕ (°)
CM	2.0	200	0.3	30
CM ~ CL	2.0	100	0.3	30
CL ~ D	2.0	50	0.4	30

Table 8.7 Cases for Crack Analysis of Tunnel Linings in Database Development

Case No.	Name	Shape	Type of Cracks	Lining Thickness t (cm)	Target CMOD (mm)
1	BC1-t20	Horseshoe Type	Type ①: SL crack	20	1, 3, 5
2	BC1-t30			30	1, 3, 5
3	BC1-t50			50	1, 3, 5
4	BC2-t20		Type ②: Arch crack	20	1, 3, 5
5	BC2-t30			30	1, 3, 5
6	BC2-t50			50	1, 3, 5
7	BC3-t20		Type ③: Crown crack	20	1, 3, 5
8	BC3-t30			30	1, 3, 5
9	BC3-t50			50	1, 3, 5

Table 8.7 Cases for Crack Analysis of Tunnel Linings in Database Development—cont'd

Case No.	Name	Shape	Type of Cracks	Lining Thickness t (cm)	Target CMOD (mm)
10	HC1-t20	Calash Type	Type ①: Wall crack	20	1, 3, 5
11	HC1-t30			30	1, 3, 5
12	HC1-t50			50	1, 3, 5
13	HC2-t20		Type ②: Shoulder crack	20	1, 3, 5
14	HC2-t30			30	1, 3, 5
15	HC2-t50			50	1, 3, 5
16	HC3-t20		Type ③: Crown crack	20	1, 3, 5
17	HC3-t30			30	1, 3, 5
18	HC3-t50			50	1, 3, 5

Table 8.8 Cases for Ground Pressure Analysis in Database Development

Case No.	Cross-Sectional Shape	Depth of Loosening Zone (m)	Classification of Rock Mass
1	Horseshoe Type	1	CL ~ D
2		2	CL ~ D
3		4	CL ~ D
4		6	CL ~ D
5		8	CL ~ D
6		10	CL ~ D
7		1	CM ~ CL
8		2	CM ~ CL
9		4	CM ~ CL
10		6	CM ~ CL
11		8	CM ~ CL
12		10	CM ~ CL
13		1	CM
14		2	CM

(Continued)

Table 8.8 Cases for Ground Pressure Analysis in Database Development—cont'd

Case No.	Cross-Sectional Shape	Depth of Loosening Zone (m)	Classification of Rock Mass
15		4	CM
16		6	CM
17		8	CM
18		10	CM
19	Calash Type	1	CL ~ D
20		2	CL ~ D
21		4	CL ~ D
22		6	CL ~ D
23		8	CL ~ D
24		10	CL ~ D
25		1	CM ~ CL
26		2	CM ~ CL
27		4	CM ~ CL
28		6	CM ~ CL
29		8	CM ~ CL
30		10	CM ~ CL
31		1	CM
32		2	CM
33		4	CM
34		6	CM
35		8	CM
36		10	CM

which are combinations of the cross-sectional shape of the tunnel, the depth of loosening zone, and the classification of rocks.

Material properties of concrete and pseudoshell are assumed to be the same as those used in the previous parametric studies of the pseudoshell model, as presented in Table 8.1. Notice that the material properties of concrete are not treated parametrically because in a preliminary study it was found that by changing these properties, the relations between cross-sectional deformation and the CMOD remained basically unchanged, especially when the CMOD exceeded 0.1 mm. For crack analysis the bilinear tension-softening relation in Figure 4.9 is assumed.

8.6.2 Relations between Cross-Sectional Deformation and the CMOD

Figures 8.30 to 8.32 present the half-FE-models of the horseshoe type for Cases 1 to 9 as listed in Table 8.7. As shown, to propagate each main crack from the inside of the tunnel, a dummy load is applied on the pseudoshell along the crack path, and the region of disconnection between the lining and shell is identified.

Similarly, the half-FE-models of the calash type for Cases 10 to 18 in Table 8.7 are shown in Figures 8.33 to 8.35. Notice that in each case the wall thickness of the pseudoshell is one-fifth the lining thickness.

Crack analysis is carried out on each of the FE models, and the obtained relations between the cross-sectional deformation and the CMOD are summarized in Figure 8.36 based on the type of crack, the thickness of the lining, and the type of tunnel. For the horseshoe type, the increase of displacement at the spring line is the largest, which is followed by that at the crown. As seen, the increase of displacement in the arch is much less. On the other hand, for the calash type the increase of displacement at the crown is large and comparable to that in the wall. For details on these relations, refer to Figure 8.36.

For further reference, examples of crack propagation in the deformed lining for each crack pattern are shown in Figure 8.37 for the horseshoe type and in Figure 8.38 for the calash type (thickness of lining = 30 cm; CMOD = 3 mm). Also, examples of the relations between displacement and the thickness of the lining for each crack pattern are shown in Figure 8.39 for the horseshoe type and in Figure 8.40 for the calash type (CMOD = 3 mm).

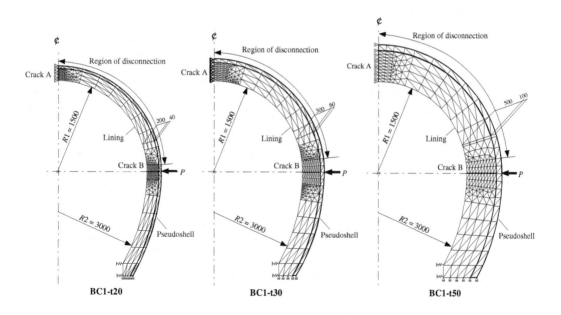

FIGURE 8.30 FE models (dimensions in mm) for the horseshoe type with SL crack (BC1).

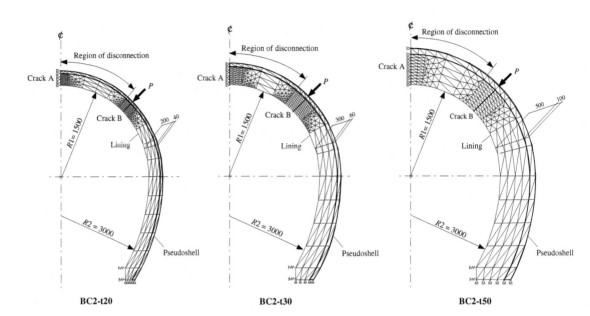

FIGURE 8.31 FE models (dimensions in mm) for the horseshoe type with arch crack (BC2).

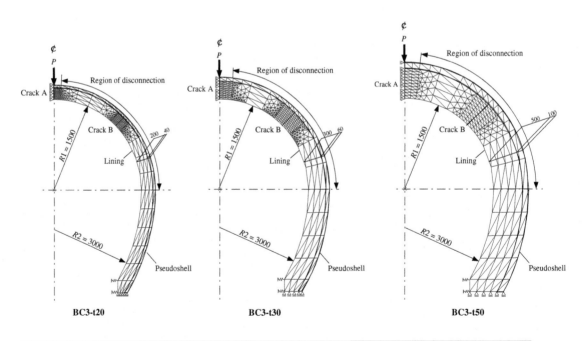

FIGURE 8.32 FE models (dimensions in mm) for the horseshoe type with crown crack (BC3).

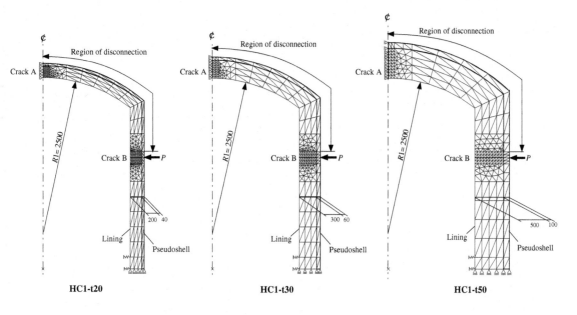

FIGURE 8.33 FE models (dimensions in mm) for the calash type with wall crack (HC1).

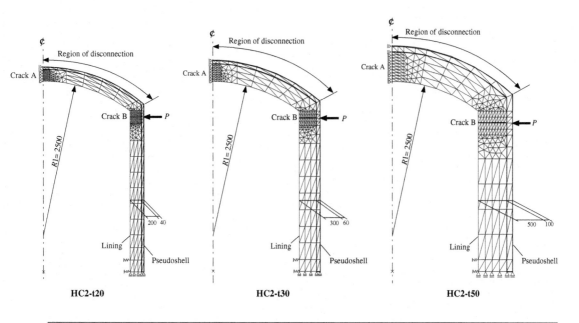

FIGURE 8.34 FE models (dimensions in mm) for the calash type with shoulder crack (HC2).

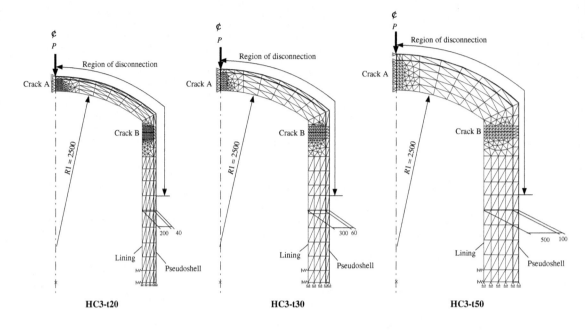

FIGURE 8.35 FE models (dimensions in mm) for the calash type with crown crack (HC3).

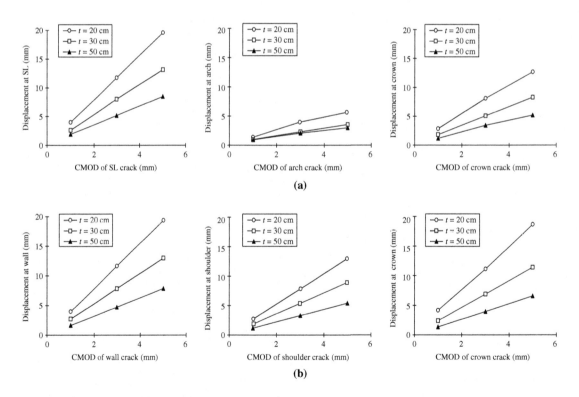

FIGURE 8.36 Cross-sectional deformation versus CMOD relations for database development: (a) horseshoe type and (b) calash type.

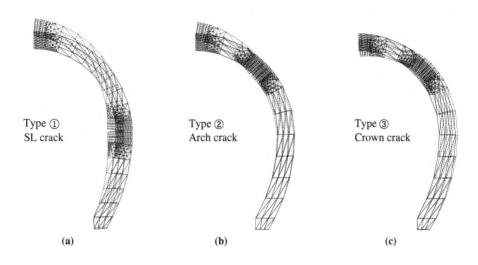

FIGURE 8.37 Lining deformation for (a) SL crack, (b) arch crack, and (c) crown crack (horseshoe type; lining thickness = 30 cm; CMOD = 3 mm).

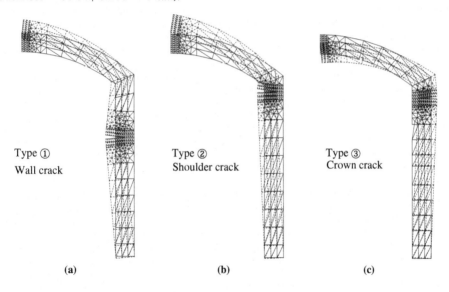

FIGURE 8.38 Lining deformation for (a) wall crack, (b) shoulder crack, and (c) crown crack (calash type; lining thickness = 30 cm; CMOD = 3 mm).

8.6.3 **Relations between Pressure Load and Cross-Sectional Deformation**

Figure 8.41 illustrates the FE models of quasi loosening zone for Cases 1 to 18 in Table 8.8 (the horseshoe type), and Figure 8.42 presents those for Cases 19 to 36 in Table 8.8 (the calash type). As shown, in each case the prescribed vertical and lateral pressure loads are applied to the model

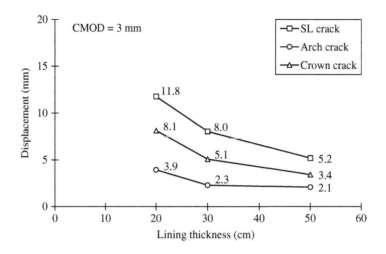

FIGURE 8.39 Cross-sectional deformation versus lining thickness relations for each crack pattern (horseshoe type).

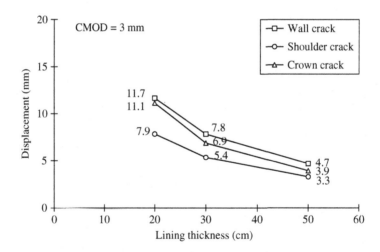

FIGURE 8.40 Cross-sectional deformation versus lining thickness relations for each crack pattern (calash type).

as the external loads, which are calculated from the depth of loosening zone and the unit weight of rocks while assuming that the coefficient of lateral pressure equals 1.0. Linear elastic analysis is then carried out to obtain the ground deformation under the given load. Based on these analyses, the relations between pressure load and cross-sectional deformation at the three crack locations are obtained for each class of rock and for each type of tunnel, as shown in Figure 8.43. As seen, these relations are basically linear.

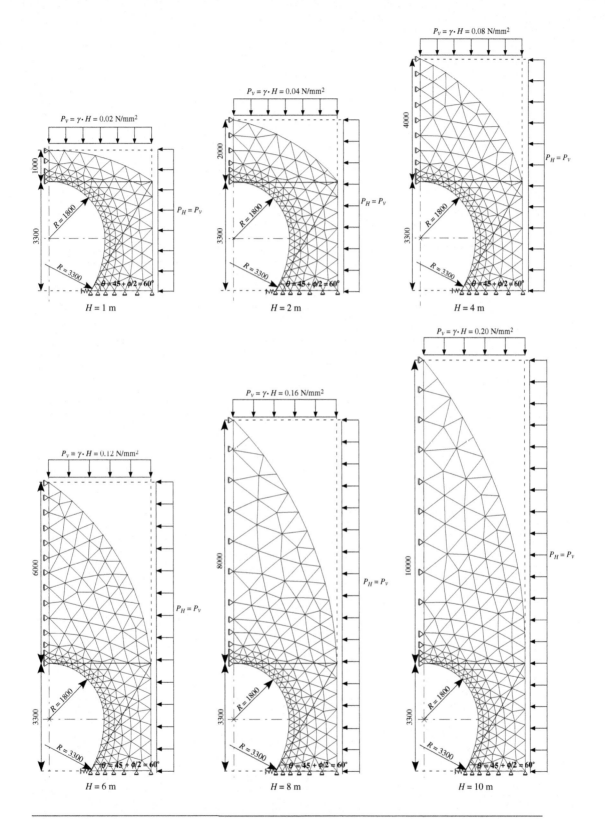

FIGURE 8.41 FE models (dimensions in mm) of quasi loosening zone with different depths (horseshoe type).

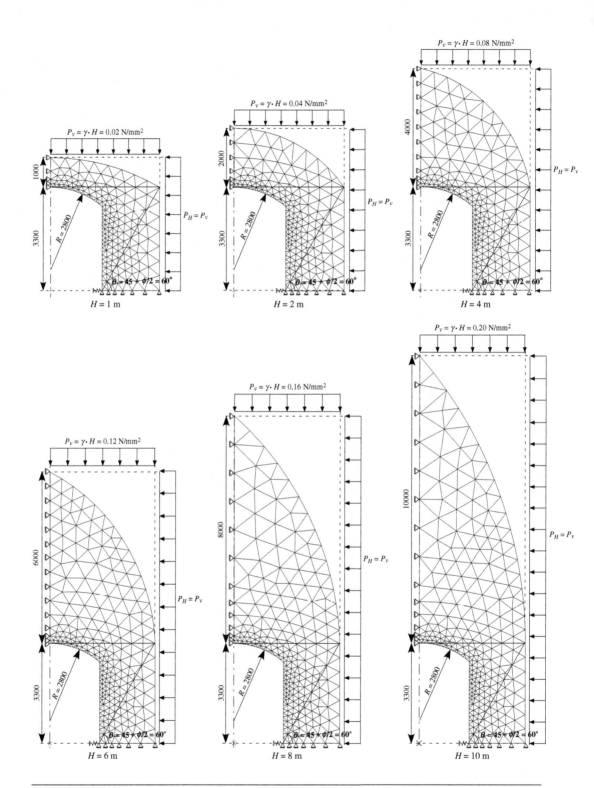

FIGURE 8.42 FE models (dimensions in mm) of quasi loosening zone with different depths (calash type).

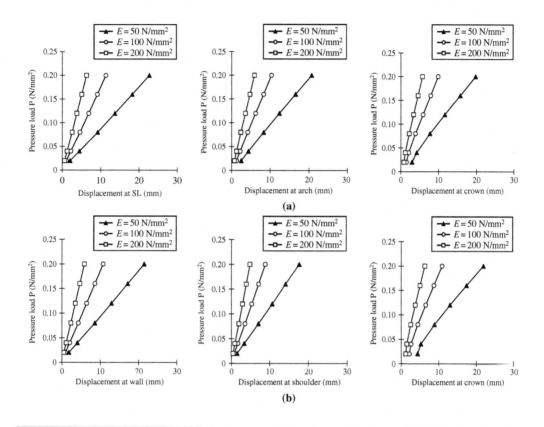

FIGURE 8.43 Pressure load versus cross-sectional deformation relations for database development: (a) horseshoe type and (b) calash type.

8.6.4 Two-Step Procedure for Determining External Loads by the CMOD and Development of Database

So far, two relationships have been established: the relations between cross-sectional deformation and the CMOD, and the relations between pressure load and cross-sectional deformation. Based on these two relations, a two-step procedure for determining the external loads from the CMOD now becomes available. As shown in Figure 8.44, this procedure requires the cross-sectional deformation at a designated crack to be determined from the first relation based on the CMOD of that crack. Then, based on the second relation, the external loads can be found by the cross-sectional deformation at the designated crack.

As an example, suppose that a crack is identified at the crown of a model tunnel and its crack width is 5 mm. Assume further that the tunnel is the horseshoe type, the thickness of lining is

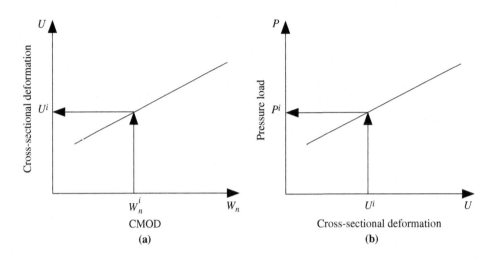

FIGURE 8.44 A two-step procedure for determining the external loads based on the CMOD: (a) to obtain deformation from CMOD and (b) to obtain load from deformation.

30 cm, and the surrounding rock mass is classified as the CL-D grade. Based on the corresponding relations in Figure 8.36, the cross-sectional deformation at the crown is then obtained as 8.3 mm. Next, from the relations between load and displacement in Figure 8.43, the ground pressure is found to be 0.086 N/mm^2. The depth of the loosening zone is then obtained as 4.3 m by dividing the pressure load by the unit weight of rocks. In this example, if the type of tunnel is changed to the calash type while keeping all the other conditions unchanged, the displacement at the crown then becomes 11.4 mm, the pressure load is 0.104 N/mm^2, and the depth of loosening zone is 5.2 m.

As another example, assume that two cracks exist symmetrically at the shoulders of a calash-type tunnel and the crack widths are approximately 1 mm. Suppose that the thickness of lining is 20 cm, and the rock mass is identified as the CM grade. Then, based on the corresponding relations in Figures 8.36 and 8.43, the displacement at the shoulder is obtained as 2.7 mm, the ground pressure is 0.092 N/mm^2, and the depth of loosening zone is 4.6 m.

Based on this two-step procedure, a database is developed for the two types of the model tunnel, as shown in Table 8.9 for the horseshoe type and in Table 8.10 for the calash type. Clearly, these tables provide a shortcut to finding out the external loads acting on an aging tunnel with cracks. By applying the same principles in developing this database, various kinds of databases can be built for different types of tunnels, which provides not only an effective means for monitoring the ground pressure for risk evaluations but also the design loads when tunnels with serious crack problems have to be reinforced.

Table 8.9 Database for Obtaining the Ground Pressure Based on the CMOD (Horseshoe Type)

Case No.	Name	Type of Crack	Thickness of Lining t (cm)	CMOD W_n (mm)	Displacement U (mm)	Classification of Rock E (N/mm^2)	Ground Pressure P_v (N/mm^2)	Depth of Loosening Zone H (m)
1	BC1-t20		20	1.0	4.0	50 (CL ~ D)	0.036	1.8
						100 (CM ~ CL)	0.070	3.5
						200 (CM)	0.126	6.3
				3.0	11.8	50 (CL ~ D)	0.104	5.2
						100 (CM ~ CL)	0.208	10.4
						200 (CM)	0.378	18.9
				5.0	19.6	50 (CL ~ D)	0.172	8.6
						100 (CM ~ CL)	0.350	17.5
		Type ①: SL crack				200 (CM)	0.632	31.6
2	BC1-t30		30	1.0	2.6	50 (CL ~ D)	0.024	1.2
						100 (CM ~ CL)	0.046	2.3
						200 (CM)	0.082	4.1
				3.0	8.0	50 (CL ~ D)	0.070	3.5
						100 (CM ~ CL)	0.140	7.0
						200 (CM)	0.256	12.8
				5.0	13.2	50 (CL ~ D)	0.116	5.8
						100 (CM ~ CL)	0.234	11.7
						200 (CM)	0.424	21.2

(Continued)

Table 8.9 Database for Obtaining the Ground Pressure Based on the CMOD (Horseshoe Type)—cont'd

Case No.	Name	Type of Crack	Thickness of Lining t (cm)	CMOD W_n (mm)	Displacement U (mm)	Classification of Rock E (N/mm²)	Ground Pressure P_v (N/mm²)	Depth of Loosening Zone H (m)
3	BC1-t50		50	1.0	1.9	50 (CL ~ D)	0.018	0.9
						100 (CM ~ CL)	0.034	1.7
						200 (CM)	0.060	3.0
				3.0	5.2	50 (CL ~ D)	0.046	2.3
						100 (CM ~ CL)	0.090	4.5
						200 (CM)	0.164	8.2
				5.0	8.5	50 (CL ~ D)	0.074	3.7
						100 (CM ~ CL)	0.150	7.5
						200 (CM)	0.272	13.6
4	BC2-t20		20	1.0	1.4	50 (CL ~ D)	0.008	0.4
						100 (CM ~ CL)	0.024	1.2
						200 (CM)	0.046	2.3
				3.0	3.9	50 (CL ~ D)	0.036	1.8
						100 (CM ~ CL)	0.074	3.7
						200 (CM)	0.132	6.6
				5.0	5.6	50 (CL ~ D)	0.052	2.6
						100 (CM ~ CL)	0.108	5.4
						200 (CM)	0.190	9.5

Type ②: Arch crack

#	Specimen						
5	BC2-t30	30	1.0	1.0	50 (CL ~ D)	0.004	0.2
					100 (CM ~ CL)	0.016	0.8
					200 (CM)	0.032	1.6
			3.0	2.3	50 (CL ~ D)	0.018	0.9
					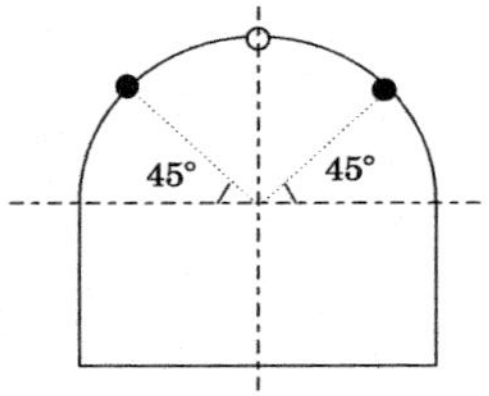 100 (CM ~ CL)	0.042	2.1
					200 (CM)	0.078	3.9
			5.0	3.5	50 (CL ~ D)	0.032	1.6
					100 (CM ~ CL)	0.066	3.3
					200 (CM)	0.118	5.9
6	BC2-t50	50	1.0	0.9	50 (CL ~ D)	0.004	0.2
					100 (CM ~ CL)	0.014	0.7
					200 (CM)	0.028	1.4
			3.0	2.1	50 (CL ~ D)	0.016	0.8
					100 (CM ~ CL)	0.038	1.9
					200 (CM)	0.070	3.5
			5.0	3.0	50 (CL ~ D)	0.026	1.3
					100 (CM ~ CL)	0.056	2.8
					200 (CM)	0.102	5.1

(Continued)

261

Table 8.9 Database for Obtaining the Ground Pressure Based on the CMOD (Horseshoe Type)—cont'd

Case No.	Name	Type of Crack	Thickness of Lining t (cm)	CMOD W_n (mm)	Displacement U (mm)	Classification of Rock E (N/mm²)	Ground Pressure P_v (N/mm²)	Depth of Loosening Zone H (m)
7	BC3-t20		20	1.0	2.8	50 (CL ~ D)	0.018	0.9
						100 (CM ~ CL)	0.056	2.8
						200 (CM)	0.100	5.0
				3.0	8.1	50 (CL ~ D)	0.084	4.2
						100 (CM ~ CL)	0.164	8.2
						200 (CM)	0.284	14.2
				5.0	12.7	50 (CL ~ D)	0.130	6.5
						100 (CM ~ CL)	0.258	12.9
						200 (CM)	0.444	22.2
8	BC3-t30	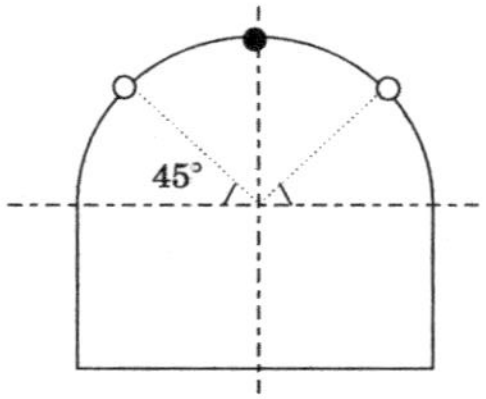	30	1.0	1.8	50 (CL ~ D)	0.002	0.1
						100 (CM ~ CL)	0.030	1.5
						200 (CM)	0.064	3.2
				3.0	5.1	50 (CL ~ D)	0.050	2.5
						100 (CM ~ CL)	0.106	5.3
						200 (CM)	0.180	9.0
				5.0	8.3	50 (CL ~ D)	0.086	4.3
						100 (CM ~ CL)	0.168	8.4
						200 (CM)	0.292	14.6

Type ③: Crown crack

45°

9	BC3-t50	50	1.0	1.2	50 (CL ~ D)	0.001	0.1
					100 (CM ~ CL)	0.012	0.6
					200 (CM)	0.042	2.1
			3.0	3.4	50 (CL ~ D)	0.028	1.4
					100 (CM ~ CL)	0.070	3.5
					200 (CM)	0.122	6.1
			5.0	5.2	50 (CL ~ D)	0.052	2.6
					100 (CM ~ CL)	0.108	5.4
					200 (CM)	0.184	9.2

Table 8.10 Database for Obtaining the Ground Pressure Based on the CMOD (Calash Type)

Case No.	Name	Type of Crack	Thickness of Lining t (cm)	CMOD W_n (mm)	Displacement U (mm)	Classification of Rock E (N/mm²)	Ground Pressure P_v (N/mm²)	Depth of Loosening Zone H (m)
10	HC1-t20		20	1.0	3.9	50 (CL ~ D)	0.046	2.3
						100 (CM ~ CL)	0.088	4.4
						200 (CM)	0.162	8.1
				3.0	11.7	50 (CL ~ D)	0.132	6.6
						100 (CM ~ CL)	0.266	13.3
						200 (CM)	0.488	24.4
				5.0	19.4	50 (CL ~ D)	0.220	11.0
						100 (CM ~ CL)	0.442	22.1
						200 (CM)	0.810	40.5
11	HC1-t30		30	1.0	2.7	50 (CL ~ D)	0.034	1.7
						100 (CM ~ CL)	0.062	3.1
						200 (CM)	0.112	5.6
				3.0	7.8	50 (CL ~ D)	0.088	4.4
						100 (CM ~ CL)	0.178	8.9
						200 (CM)	0.326	16.3
				5.0	13.0	50 (CL ~ D)	0.148	7.4
						100 (CM ~ CL)	0.296	14.8
						200 (CM)	0.542	27.1

Type ①: Wall crack

12	HC1-t50	50	1.0	1.6	50 (CL ~ D)	0.022	1.1
					100 (CM ~ CL)	0.040	2.0
					200 (CM)	0.068	3.4
			3.0	4.7	50 (CL ~ D)	0.056	2.8
					100 (CM ~ CL)	0.106	5.3
					200 (CM)	0.196	9.8
			5.0	7.9	50 (CL ~ D)	0.090	4.5
					100 (CM ~ CL)	0.180	9.0
					200 (CM)	0.330	16.5
13	HC2-t20	20	1.0	2.7	50 (CL ~ D)	0.028	1.4
					100 (CM ~ CL)	0.052	2.6
					200 (CM)	0.092	4.6
			3.0	7.9	50 (CL ~ D)	0.074	3.7
					100 (CM ~ CL)	0.148	7.4
					200 (CM)	0.272	13.6
			5.0	13.0	50 (CL ~ D)	0.122	6.1
					100 (CM ~ CL)	0.244	12.2
					200 (CM)	0.448	22.4

(Continued)

Table 8.10 Database for Obtaining the Ground Pressure Based on the CMOD (Calash Type)—cont'd

Case No.	Name	Type of Crack	Thickness of Lining t (cm)	CMOD W_n (mm)	Displacement U (mm)	Classification of Rock E (N/mm²)	Ground Pressure P_v (N/mm²)	Depth of Loosening Zone H (m)
14	HC2-t30	Type ②: Shoulder crack	30	1.0	1.9	50 (CL ~ D)	0.020	1.0
						100 (CM ~ CL)	0.038	1.9
						200 (CM)	0.066	3.3
				3.0	5.4	50 (CL ~ D)	0.052	2.6
						100 (CM ~ CL)	0.102	5.1
						200 (CM)	0.186	9.3
				5.0	8.9	50 (CL ~ D)	0.084	4.2
						100 (CM ~ CL)	0.166	8.3
						200 (CM)	0.306	15.3
15	HC2-t50		50	1.0	1.2	50 (CL ~ D)	0.014	0.7
						100 (CM ~ CL)	0.026	1.3
						200 (CM)	0.044	2.2
				3.0	3.3	50 (CL ~ D)	0.034	1.7
						100 (CM ~ CL)	0.062	3.1
						200 (CM)	0.114	5.7
				5.0	5.4	50 (CL ~ D)	0.052	2.6
						100 (CM ~ CL)	0.102	5.1
						200 (CM)	0.186	9.3

No.	Specimen						
16	HC3-t20	20	1.0		50 (CL ~ D)	0.016	0.8
				4.2	100 (CM ~ CL)	0.076	3.8
					200 (CM)	0.134	6.7
			3.0		50 (CL ~ D)	0.102	5.1
				11.1	100 (CM ~ CL)	0.204	10.2
					200 (CM)	0.352	17.6
			5.0		50 (CL ~ D)	0.172	8.6
				18.7	100 (CM ~ CL)	0.340	17.0
					200 (CM)	0.594	29.7
17	HC3-t30	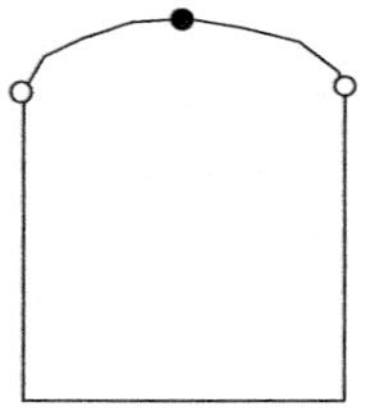 30	1.0		50 (CL ~ D)	0.005	0.2
				2.4	100 (CM ~ CL)	0.030	1.5
					200 (CM)	0.074	3.7
			3.0		50 (CL ~ D)	0.058	2.9
				6.9	100 (CM ~ CL)	0.126	6.3
					200 (CM)	0.220	11.0
			5.0		50 (CL ~ D)	0.104	5.2
				11.4	100 (CM ~ CL)	0.208	10.4
					200 (CM)	0.362	18.1

Type ③: Crown crack

(Continued)

267

Table 8.10 Database for Obtaining the Ground Pressure Based on the CMOD (Calash Type)—cont'd

Case No.	Name	Type of Crack	Thickness of Lining t (cm)	CMOD W_n (mm)	Displacement U (mm)	Classification of Rock E (N/mm²)	Ground Pressure P_v (N/mm²)	Depth of Loosening Zone H (m)
18	HC3-t50		50	1.0	1.4	50 (CL ~ D)	0.003	0.1
						100 (CM ~ CL)	0.018	0.9
						200 (CM)	0.034	1.7
				3.0	3.9	50 (CL ~ D)	0.008	0.4
						100 (CM ~ CL)	0.068	3.4
						200 (CM)	0.124	6.2
				5.0	6.6	50 (CL ~ D)	0.056	2.8
						100 (CM ~ CL)	0.122	6.1
						200 (CM)	0.210	10.5

REFERENCES

Adachi, T. and Oka, F. (1992). "An elasto-plastic constitutive model for soft rock with strain softening." *J. Geotechnical Engineering*, 445(18), 9–16.

Adachi, T., Oka, F., Soraoka, H., and Koike, M. (1998). "Time dependent behavior of soft rock and its essential interpretation." *J. Geotechnical Engineering*, 596(43), 1–10.

Jiang, Y., Esaki, T., and Yokota, Y. (1993). "Stability analysis and tunnel design for soft rock ground." *Proc. Tunnel Engineering*, 3, 17–24.

Kitagawa, T. and Inagaki, D. (1993). "Deformational behavior of surrounding soft rock mass in tunneling." *J. Geotechnical Engineering*, 463(22), 105–114.

Tasaka, Y., Uno, H., Ohmori, T., and Kudoh, K. (2000). "A joint and rock failure strain-softening model and its application to the excavation simulation of large-scale underground caverns." *J. Geotechnical Engineering*, 652(51), 73–90.

Terzaghi, X. (1959). *Theoretical Soil Mechanics*. John Wiley & Sons.

Yin, J., Wu, Z., Asakusa, T., and Ota, H. (2001). "Cracking and failure behavior of concrete tunnel lining predicted by smeared crack model." *J. Structural Mechanics/Earthquake Engineering*, 668(54), 17–27.

Zhang, F. (1994). "Constitutive models for geologic materials and their application to excavation problems." PhD Thesis, Kyoto University.

Computer Program for Mode-1 Type Crack Analysis in Concrete Using EFCM (CAIC-M1.FOR)

9.1 OVERVIEW OF THE PROGRAM

This program performs the mode-I type crack analysis in structural concrete based on the Extended Fictitious Crack Model (EFCM). The main features of this program are as follows:

Scope of analysis: Numerical analysis of concrete fracture involving an arbitrary number of discrete cracks of the mode-I type

Numerical method: Two-dimensional finite element method

Type of finite element: Three nodes, triangular elements

Theory of crack analysis: Extended fictitious crack model

Precision of calculation: Double precision

Matrix solution method: Gaussian elimination

Theory of stress analysis: Modified incremental stress analysis method; refer to Chapter 3.

Crack path modeling: Duplicate nodes on element boundaries; refer to Chapter 3.

Crack path correction: Automatic path-shifting method; refer to Chapter 3.

Applicable load types: Constant loads such as self-weight, distributed loads, and concentrated loads

Output: Crack propagation pattern, load, element stress, nodal displacement, and nodal force; the universal format can be chosen for the output of element stress and nodal displacement.

Notice that this program was written to carry out crack analysis in concrete strictly following the theory of the extended fictitious crack model; programming elegance was not a priority. The documented subroutines are listed in Table 9.1.

9.2 STRUCTURE OF THE PROGRAM

The structure of the program is illustrated in four flowcharts in Figures 9.1 to 9.4. Each chart presents the computational procedure in one of the four blocks that constitute the program. Notice that in these charts, the names of the subroutines used in the program are shown in boldface type.

Table 9.1 Documented Subroutines

Name of Subroutine	Main Function
SETPAR	Reading parameters on dimensions
MAINCN	Performing core computations in the EFCM
INPUT	Reading input data
GAUSSQ	Setting integration constants
INITL	Initializing variables (1)
ZERO	Initializing variables (2)
TFORCE	Determining the true crack propagation pattern
SETBND	Setting crack boundary conditions
EFFECT	Calculating influence coefficients and solving crack equations
TOPDWN	Judging the validity of a solution of the crack equation and modifying crack boundary conditions accordingly
STSTEP	Storing crack propagation patterns and solutions of crack equations
SETLOD	Setting the external load and cohesive force for the true crack propagation pattern
LOADPS	※ Setting equivalent nodal loads for concentrated and distributed loads
NLOAD	※ Setting equivalent nodal loads for constant loads
COEFF	Setting unit cohesive force
STIFFP	※ Evaluating element stiffness matrix
ASSEMB	※ Superimposing nodal loads and stiffness matrices
GREDUC	※ Gaussian elimination (forward elimination)
BAKSUB	※ Gaussian elimination (backward substitution)
MISAM	Stress analysis by the modified incremental stress analysis method
RESIDU	※ Calculating equivalent nodal forces
CONVER	Convergence judgment
FNTIP	Calculating nodal force at the crack tip
WNCOD	Calculating crack-opening displacement at the crack node
CRACK	Solving crack equations
OVERCK	Evaluating the tip stresses at subcracks
OUTPUT	Output of numerical results (1)

Table 9.1 Documented Subroutines—cont'd

Name of Subroutine	Main Function
UNVOUT	Output of numerical results (2) (universal format)
CRKDR	Determining path correction methods and modifying crack paths
RMESH1	Correcting the crack path (to left from the initial path)
RMESH2	Correcting the crack path (to right from the initial path)
MODFT	Redefining the limit nodal force (after path correction)
BMATPS	✳ Evaluating B matrix (strain-displacement relationship)
MODPS	Evaluating D matrix (stress-strain relationship)
DBE	✳ Formulating the product of matrices, DB
SFR2	Defining shape functions for triangular elements and their derivatives
JACOB2	✳ Evaluating the Jacobian matrix
LINEAR	✳ Calculating stress and displacement through linear elastic analysis
MINVS	Inverse matrix calculation

Note: The subroutines marked with an asterisk (*) were developed with reference to the program written by Owen and Hinton (1980).

Figure 9.1 presents the main flowchart of the program, which is composed of input/output data processing, crack pattern analysis, stress analysis, and crack path correction. The flowchart for the crack pattern analysis block is shown in Figure 9.2, which is the core of the crack analysis based on the EFCM. The functions of the block include searching for the solutions of various crack equations and determining the true crack propagation pattern or cracking mode. After setting the external loads and crack boundary conditions for the true crack pattern, the block transfers the processing to the stress analysis block.

The flowchart for formulating and solving a crack equation is shown in Figure 9.3. This block is located in TFORCE, and a crack equation is formulated by assuming a single-active cracking mode. Figure 9.4 presents the flowchart for stress analysis. In consideration of a possible extension to elastoplastic analysis, a modified incremental stress analysis method (MISAM) is employed. In the case of nonlinear stress analysis, modifications of the subroutine "RESIDU," shown in Figure 9.4, are required, which can be made with relative ease with reference to the program written by Owen and Hinton (1980). This block is used to carry out stress analysis after the determination of the true crack propagation pattern. In formulating a crack equation, it is also used to calculate the influence coefficients with unit external load, unit cohesive force, and the constant load. The subroutine MISAM, which is based on the method of the modified incremental stress analysis, is not necessary for calculating influence coefficients.

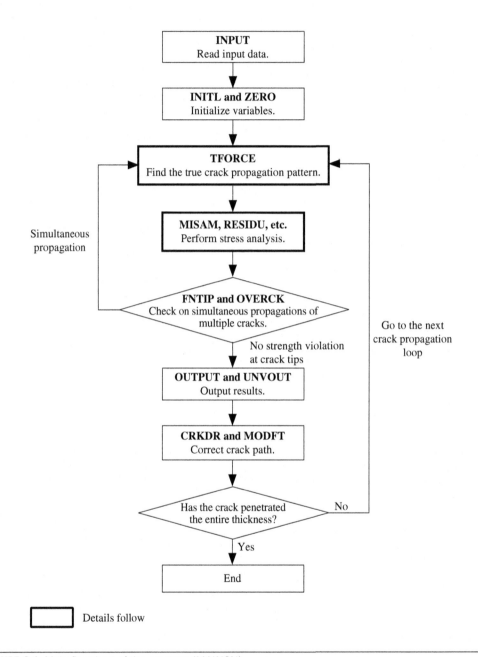

FIGURE 9.1 Main flowchart of the program (MAINCN).

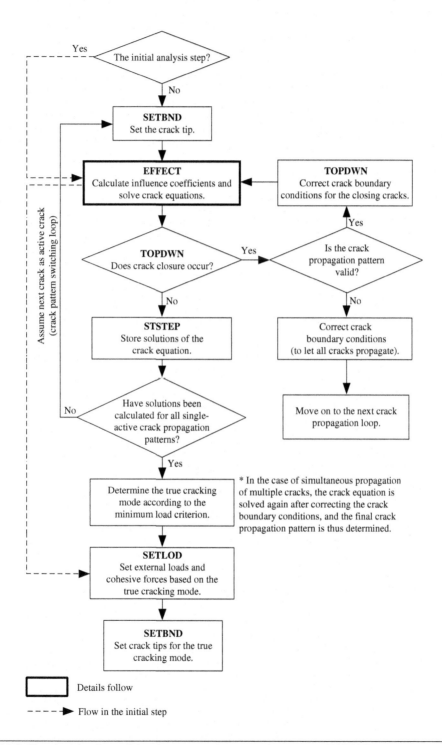

FIGURE 9.2 Flowchart for determining the true crack propagation pattern (TFORCE).

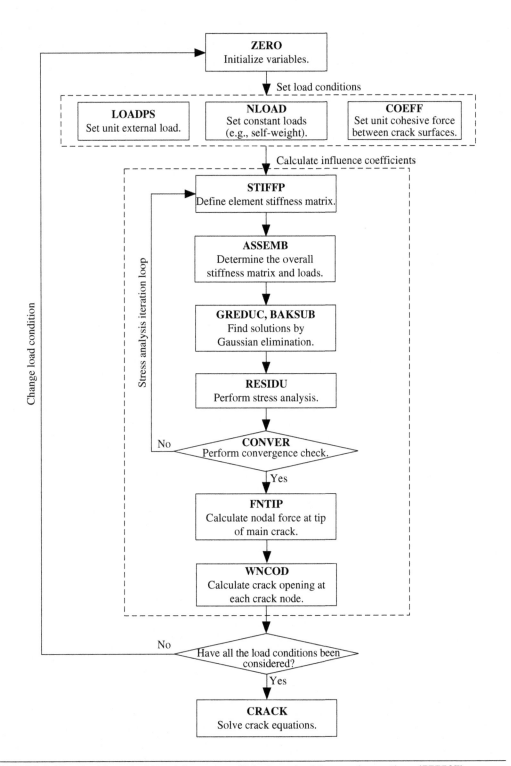

FIGURE 9.3 Flowchart for calculating influence coefficients and solving crack equations (EFFECT).

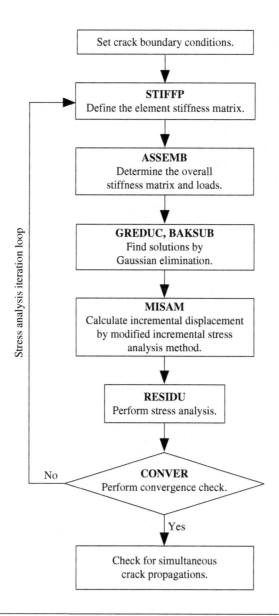

FIGURE 9.4 Flowchart for stress analysis.

9.3 MAIN RULES

Main rules and issues related to the programming are discussed in this section.

Configuration of Triangular Elements

The triangular elements used in the program must be connected in such a manner that a lower left element and an upper right element adjoin each other relative to the crack propagation direction. The configuration of nodes in this case is shown in Figure 9.5.

Crack Path Modeling

Two Modeling Methods: A crack path can be defined in two ways, depending on the type of crack concerned (discriminated by means of the parameter IFCRK): defining a crack on a model boundary and defining a crack at any location inside the model. In the case of the first way, crack propagation is modeled by removing the displacement constraints at the boundary. If a crack is defined in this manner, the crack path is then fixed, and no path correction is carried out. In the case of the second way, it is necessary in advance to define duplicate nodes and dummy elements that are formed by these nodes. As shown in Figure 9.6, the configuration of dummy elements is the same as that of other triangular elements. Of the three nodes, however, nodes number 1 and number 3 are duplicate nodes interconnected by internal springs that possess spring moduli of "∞ equivalent" and zero and provide two types of connection: rigid connection or free. Crack opening and closing are modeled through these internal springs. For a dummy element, its stiffness matrix is evaluated and superposed on the overall stiffness matrix, but stress analysis is not performed.

Crack Surface and Crack Nodes: The numbers assigned to a pair of duplicate nodes or dual nodes on the crack surface are as follows. When the direction of crack propagation is represented by a local y-axis, the node on the left-hand side is defined as the "*i*-tip," and the node on the right-hand side is defined as the "*j*-tip" (Figure 9.6). The nodes that constitute a crack surface are called "crack nodes."

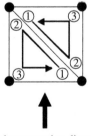

Crack propagation direction

FIGURE 9.5 Configuration of triangular elements.

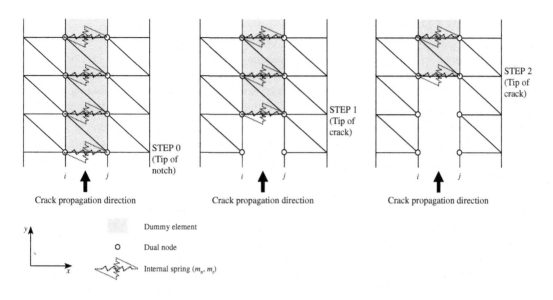

FIGURE 9.6 Defining a crack surface.

Nodal Force

To introduce the tension-softening relation of concrete, cohesive forces are applied directly on the crack surface (fictitious crack). The limit nodal force (QF1) is calculated by multiplying the tensile strength by the area of influence, which is the sum of one-half the length of the element on each side of the node multiplied by the thickness of the member. In this program, the tip of a notch is treated in a similar manner. To check the stress state at the tip of a crack or notch, the nodal force is also used.

Main Crack and Subcrack

A main crack is the crack for which a crack equation is formulated based on the assumption of a single-active crack propagation mode. The remaining cracks are assumed to be inactive in the crack equation and are called "subcracks."

Validity of Solutions

The solution of a crack equation is regarded as invalid if it is unrealistic, such as when the crack opening width or the cohesive force at a subcrack node becomes negative. If such an invalid solution has been obtained, the crack propagation pattern is then modified to close the tip of the corresponding subcrack, and the crack equation is resolved. The validity of solutions is judged by TOPDWN in TFORCE.

A solution is also regarded as invalid when the obtained external load becomes negative or when the opening displacement at any of the crack nodes of the main crack becomes negative. In such cases, the corresponding crack pattern is regarded as unrealistic, and the external load is deliberately set as infinity (∞).

Multiple-Crack Analysis

This program selects a true crack propagation pattern for N cracks from a total of N types of single-active crack modes, according to the minimum load criterion. The single-cracking mode is obtained by assuming each crack individually as an active crack with the others remaining inactive, and a set of crack equations is formulated accordingly. In the solution-finding process, the possibilities of simultaneous propagation and closure of subcracks are examined in MAINCN and TFORCE. Related programming issues are described as follows.

Single-Active Crack Propagation: TFORCE performs stress analysis after determining a true cracking mode. If the tip stress at subcracks has not reached the tensile strength, crack propagation is due to a single-active crack.

Crack Closure: After EFFECT finds solutions to the crack equation, TOPDWN evaluates possible crack closure at subcracks. Crack closure is judged to occur if the crack opening displacement or the cohesive force at the crack node immediately preceding the tip of a subcrack assumes a negative value or if the cohesive force at any of the nodes of a subcrack becomes negative. To close that crack, the duplicate nodes next to the crack tip are reconnected, and the modified crack equation is resolved. The irreversibility of material damage at the tip of a closing sub-crack is maintained by setting the cohesive stress transmitted thus far as the transient tensile strength, with which the limit nodal force is calculated.

After determining the true crack pattern, MISAM corrects the overall nodal displacement at the closed crack nodes to the same value, and the displacement thus determined is used as the initial predictor in the subsequent iterative stress analysis.

Crack Arrest: If a crack that began to propagate has stopped propagating at a subsequent step and neither continues to propagate nor closes, then the crack boundary conditions are not updated, assuming that the crack remains stationary.

Simultaneous Propagation of Multiple Cracks: TFORCE selects the true crack propagation pattern from all of the valid patterns determined earlier, according to the minimum load criterion. Based on the results of stress analysis, OVRCK checks the state of stress at the tips of all subcracks. If the tensile strength is exceeded, TFORCE is called back to modify the crack propagation pattern, assuming that multiple cracks propagate simultaneously.

In determining the simultaneous propagation of a subcrack, the following variable is defined in OVRCK:

- *Name of variable:* AAA.
- *Description:* The ratio between the nodal force at the tip of a subcrack and the limit nodal force.

Since a nodal force depends on the element length, it does not necessarily represent the exact stress at the nodal point. To avoid oscillation of solutions between crack propagation and crack closure, a strength tolerance of several percent is adopted in calculating the nodal force so the simultaneous crack propagation only occurs when the nodal force at the tip of a subcrack surely exceeds the limit nodal force. If, however, oscillation of solutions still occurs even with such a strength tolerance, the program skips the present step and moves on to the next step of analysis, assuming that no valid crack propagation pattern can be found under the current conditions (for details, see the next section).

Example

AAA = –1.00 → A simultaneous propagation occurs when the nodal force at the tip of a subcrack has reached the tensile strength.

AAA = –1.01 → A simultaneous propagation occurs when the nodal force at the tip of a subcrack has reached 1.01 times the tensile strength.

Validity of Crack Propagation Patterns: In principle, the validity of a crack propagation pattern is judged by the cracking behaviors at subcracks in MAINCN and TFORCE. If a subcrack is initially judged to propagate with the main crack, and upon solving the modified crack equation its crack opening displacement becomes negative, then the program moves on to the next stage of analysis, assuming that no valid crack pattern exists at the present computational step. When the program enters the next stage of computation, all the previous propagating cracks are automatically advanced by one step; from there, a new round of crack analysis begins. In other words, if a valid crack propagation pattern cannot be determined, the solution-finding process simply skips that step and begins the next step of crack analysis.

Also, if the main crack is found to close or the external load becomes negative when solving the crack equation for multiple crack propagation, the program takes similar measures as those just described, judging that no valid crack propagation pattern can be found in the present step. These are rare occasions, however, that occur with a considerable change of stress field, such as when using the pseudoshell model discussed in Chapter 8.

True Crack Pattern: Of all the valid crack propagation patterns obtained by solving various crack equations, one that has been selected according to the minimum load criterion is regarded temporarily as the true crack pattern. If multiple cracks are found after stress analysis to propagate simultaneously, this true crack pattern is modified accordingly. After resolving the crack equation and verifying the validity of the solutions, the final crack pattern is determined.

Main Flags

This section lists the main flags used in the program and their functions.

IR:
- The initial setting in MAINCN is IR = 0. Based on the results of stress analysis, if the nodal force at the tip of a subcrack exceeds the tensile strength in OVERCK, IR = 999 is passed to TFORCE.
- In TFORCE, of a total of NNCRK cracks, the IRth crack is defined as the main crack.
- In TFORCE, when a true solution has been determined, IR = 999 is passed to SETBND.

NNMM:
- The number assigned to the main crack in the final crack propagation pattern.
- If no valid crack propagation pattern is obtained, NNMM = 999 (TOPDWN).

NSTOP: Flag of Solution Validity:

    0: Valid solution

    1: The crack opening displacement or the cohesive force at the node preceding the tip of a sub-crack is negative

    11: The cohesive force at any node of a subcrack is negative

    22: The crack opening displacement or the cohesive force at the node preceding the tip of the main crack is negative

IFLA2: Flag of Load Setting:

    0: Unit external load

    1: The external load in the initial state of computation

    2: The external load in the subsequent steps of computation

NOVER: This flag indicates whether the nodal force at the tip of a subcrack exceeds the strength.

    999: Exceeded

    0: Not exceeded

NPARA: Flag of Path Correction:

    0: Minor correction (modifying the coordinates of nodes)

    1: Major correction—to the left of the initial path (modifying the coordinates of nodes, crack node configuration, element configuration, and nodal displacement)

    −1: Major correction—to the right of the initial path (modifying the coordinates of nodes, crack node configuration, element configuration, and nodal displacement)

Key Variables Indicating the State of Crack Propagation

In searching for a true crack propagation pattern, a number of variables are used to store the node numbers at and near the crack-tip to indicate the state of crack propagation. Main variables and their numbering rules are shown in Table 9.2 and Figure 9.7.

Table 9.2 Major Variables Related to Cracks	
NNCRK	Number of cracks
MMDP	Maximum value of the total number of crack nodes at each crack, counted from the tip of the notch
MDP(NNCRK)	Number of crack nodes at each crack, counted from the tip of the notch
MOP(MMDP+1,2, NNCRK)	Node number counted from the node immediately preceding the notch tip
NCRAK(NNCRK)	Indicating the tip of each crack [initial state: NCRAK(NNCRK) = 0]; used in MAINCN
NR(NNCRK)	Indicating the tip of each crack [initial state: NR(NNCRK) = 0]; used in TFORCE
NINTS	Total number of internal springs
MMSTP(NINTS)	Internal spring number

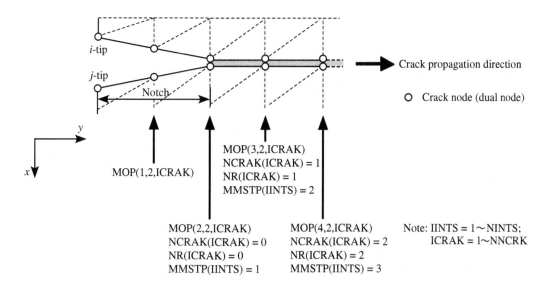

FIGURE 9.7 Numbering rules for crack nodes.

Formulation of Crack Equation

A crack equation is formulated and solved by EFFECT. By using the illustrative crack problem shown in Figure 4.5, this section explains the formulation of the crack equation in matrix forms.

Conditions of the problem (Figure 4.5):
  (a) Fracture mode: Mode I
  (b) Number of cracks: 2 (crack A and crack B)
  (c) State of crack propagation (number of crack nodes):
      Crack A: crack node $i = 1, \ldots, n$ [NR(A) $= n$ in TFORCE]
      Crack B: crack node $j = 1, \ldots, m$ [NR(B) $= m$ in TFORCE]
  (d) Main crack: Crack A
Notice that physical quantities related to an inactive crack (crack B) are denoted with asterisks.

In the crack equation, a tensile force (force that opens a crack) is defined as negative. The crack equation is expressed in the matrix form as

$$[SD] = [SE] \times [SF] \qquad (9.1)$$

with

$$[SD]^T = [BKN_a \ Q_a \ BKN_b \ Q_b \ -(Q_{la} + CRN_a)] \qquad (9.2)$$

$$[SE] = \begin{bmatrix} E_n & -AK_{aa} & 0 & -AK_{ab} & -BK_a \\ S_a & tE_n & 0 & 0 & 0 \\ 0 & -AK_{ba} & E_m & -AK_{bb} & -BK_b \\ 0 & 0 & S_b & tE_m & 0 \\ 0 & CI_{aa} & 0 & CI_{ab} & CR_a \end{bmatrix} \qquad (9.3)$$

$$[SF]^T = \begin{bmatrix} W_a & F_a & W_b^* & F_b^* & P_a \end{bmatrix} \qquad (9.4)$$

$$AK_{aa} = \begin{bmatrix} AK_{aa}^{11} & AK_{aa}^{12} & \cdot & \cdot & AK_{aa}^{1n} \\ AK_{aa}^{21} & AK_{aa}^{22} & \cdot & \cdot & AK_{aa}^{2n} \\ \cdot & & \cdot & & \cdot \\ \cdot & & & \cdot & \cdot \\ AK_{aa}^{n1} & AK_{aa}^{n2} & \cdot & \cdot & AK_{aa}^{nn} \end{bmatrix} \qquad (9.5)$$

Matrix of the crack-opening displacement at each node of crack A due to unit cohesive force at the nodes of crack A $(n \times n)$

$$AK_{ab} = \begin{bmatrix} AK_{ab}^{11} & AK_{ab}^{12} & \cdot & \cdot & AK_{ab}^{1m} \\ AK_{ab}^{21} & AK_{ab}^{22} & \cdot & \cdot & AK_{ab}^{2m} \\ \cdot & & \cdot & & \cdot \\ \cdot & & & \cdot & \cdot \\ AK_{ab}^{n1} & AK_{ab}^{n2} & \cdot & \cdot & AK_{ab}^{nm} \end{bmatrix} \qquad (9.6)$$

Matrix of the crack-opening displacement at each node of crack A due to unit cohesive force at the nodes of crack B $(n \times m)$ where subscripts a and b represent crack A and crack B, respectively. The definitions of the matrices and variables are given in the following list.

[SD]—Vector of strength properties: In the absence of constant loads, this column vector reflects the properties of the tension-softening model and the limit nodal force.
  BKN_a and BKN_b: Vector of the crack opening displacement due to constant loads at each node of crack A and crack B, respectively (column vectors of order n and order m)
  Q_a and Q_b: Vector of the Y-intercept in the tension-softening model at each node of crack A and crack B, respectively (column vectors of order n and order m)
  Q_{la}: The limit nodal force at the tip of crack A
  CRN_a: The nodal force component at the tip of crack A due to constant loads
[SE]—Influence coefficient matrix:
  E_n and E_m: Unit matrix $(n \times n)$ and unit matrix $(m \times m)$, respectively
  S_a and S_b: Matrix of the slope in the tension-softening model corresponding to the crack opening displacement of W_a and W_b^*, respectively $[(n \times n)$ and $(m \times m)]$
  AK_{ba} and AK_{bb}: Matrix of the crack opening displacement at each node of crack B due to unit cohesive force at the nodes of crack A and crack B, respectively $[(m \times n)$ and $(m \times m)]$
  BK_a and BK_b: Vector of the crack opening displacement due to a unit external load at crack A and crack B, respectively (column vectors of order n and order m)
  CI_{aa} and CI_{ab}: Vector of the nodal force at the tip of crack A due to unit cohesive force at the nodes of crack A and crack B, respectively (row vectors of order n and order m)
  CR_a: Nodal force at the tip of crack A due to a unit external load (< 0)
  T: Member thickness

[SF]—Vector of unknowns:

W_a and W_b^*: Vector of the crack opening displacement at each node of crack A and crack B, respectively (column vectors of order n and order m)

F_a and F_b^*: Vector of the cohesive force at each node of crack A and crack B, respectively (column vectors of order n and order m)

P_a: External load required to propagate crack A

Signs in the Crack Equation

This program models the crack surface by using duplicate nodes, at which the influence coefficients of the crack equation are calculated. Duplicate nodes share the same coordinates, but they are independent nodes that constitute different elements. In constructing a crack equation, therefore, it is necessary to pay attention to the signs of the influence coefficients at the i-tip and j-tip of the duplicate nodes and make appropriate decisions as to which influence coefficient to use.

As shown in Eq. (9.1), in the crack equation CR assumes a negative value. Therefore, in formulating the crack equation, it is necessary to focus on the tip of the dual nodes at which the nodal force is negative, such as the case shown in Figure 9.8 for the given direction of crack propagation. Computations of the influence coefficients in FNTIP and WNCOD follow this principle. Based on the sign conventions adopted in the program, for a symmetric problem its right half is modeled.

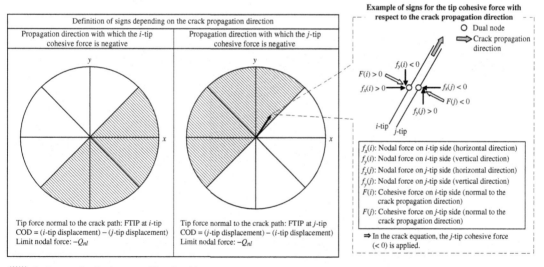

FIGURE 9.8 Definition of signs in the crack equation.

Dealing with Self-Weight and Other Constant Loads

Nodal forces at the crack tips and the crack opening displacements at crack nodes due to the self-weight and other constant loads are treated as constants and are accounted for in the vector [SD] of Eq. (9.1). Therefore, when calculating the influence coefficients under a unit external load or unit cohesive force, the state of stress due to the self-weight and other constant loads is regarded as the initial state.

Termination of Computation

This program terminates computations according to the following criteria:

- When the tip of any crack reaches the node next to the last node on the crack path that is preset across the structural member, the program terminates the calculation, judging that a complete fracture of the structural member has occurred.
- If the number of iterations in stress analysis exceeds 200, the program terminates the calculation, judging that the solution has diverged.

Crack Path Correction

Defining dummy paths for path correction: To prepare for path correction, it is necessary to define dummy paths in the vicinity of the initially defined crack path in the input data. Figure 9.9 shows a case where six dummy paths have been designated for path correction. As shown, with the given crack propagation direction, dummy paths must be defined from the left-hand side of the initial path to the right-hand side. On each dummy path, node numbers

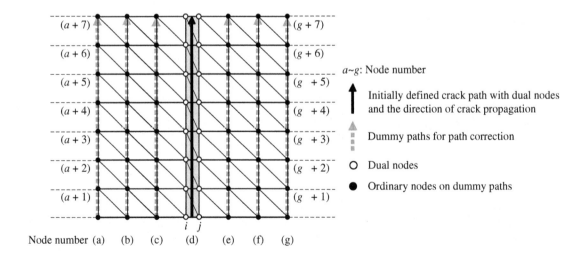

FIGURE 9.9 Designated paths for path correction and related numbering rules.

in the propagation direction must be sequential. Notice that the node number of the i-tip on the initial crack path (path d in Figure 9.9) should also be provided.

Methods of crack path correction (refer to Chapter 3): After carrying out stress analysis based on the final crack propagation pattern obtained, crack path correction is carried out in CRKDR, prior to the next step of crack analysis. Figure 9.10 illustrates the concept of path modification, along with the definitions of related variables used in the program. Figure 9.11 presents the flowchart for path correction, which contains three correction methods depending on the reference angle ϕ ($=|\theta - \alpha|/2$) and the angle γ between a new path and the initial path.

Input Data

Table 9.3 shows input data formats.

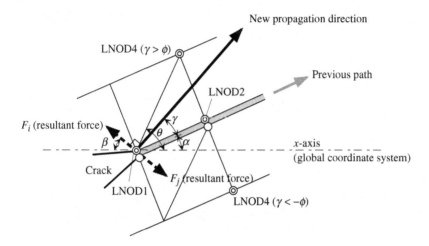

◎ Reference node for path correction

Main variables:
LNOD1: i-tip of the dual nodes at the present crack tip
LNOD2: i-tip of the next dual nodes on the previous path
LNOD4: i-tip or j-tip of the dual nodes at the crack tip of the next step propagation in the case of a major path correction
ALPHA (α): angle of the initial path from the x-axis
BETA (β): angle of the crack tip resultant from the x-axis
GAMMA (γ): angle between a new path (perpendicular to the resultant) and the initial path
THETA (θ): angle of the line segment from LNOD1 to LNOD4 from the x-axis
PHI (ϕ): reference angle for selecting a modification method in path correction ($= |\theta - \alpha|/2$)

FIGURE 9.10 Schematic illustration of path correction and major variables.

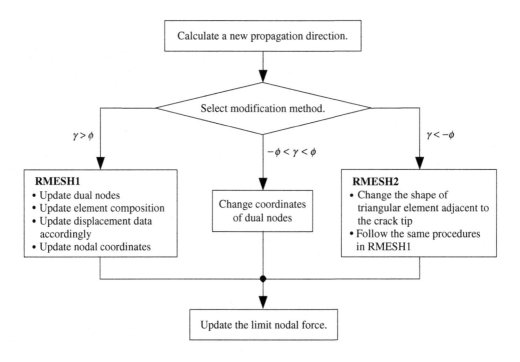

FIGURE 9.11 Flowchart for path correction in CRKDR (for path correction, see Figures 3.8–3.11).

Table 9.3 Input Data Formats

Number of Lines	Format	Variable	Description
1	A100	TITLE	Title
1	I5	NT	Total number of nodes
	I5	NE	Total number of elements
	I5	NMATS	Number of material types
	I5	NNCRK	Total number of cracks
	I5	NINTS	Number of internal springs with which the crack surface is defined
	I5	NNUP	Number of dummy paths for path correction
	I5	NTYPE	Problem-type indicator (1: plane strain; 2: plane stress)
	F10.2	THICK	Member thickness (= 0 for plane strain)
	F10.2	TOLER	Convergence tolerance (%)

Table 9.3 Input Data Formats—cont'd

Number of Lines	Format	Variable	Description
1	I5	MDP()	Total number of nodes of each crack
1	I5	NANTS()	Number of internal springs of each crack
1	I5	IFCRK()	Crack surface-type indicator (0: dual nodes; 1: half-model boundary in the y-axis; 2: half-model boundary in the x-axis)
1 × NT	Free (up to 7 columns)	INODE	Node number
	I1	INF()	Horizontal restraint conditions (0: free; 1: fixed)
	I1	INF()	Vertical restraint conditions (0: free; 1: fixed)
	F10.3	COORD(*,1)	X coordinate
	F10.3	COORD(*,2)	Y coordinate
1 × NF	I7	N	Element number
	I7	MATNO()	Material number (0: dummy element)
	3I5	NKOM(*,J)	Node number of jth node in an element ($j = 1$ to 3)
NMATS	E10.3	PROPS(*,1)	Modulus of elasticity
	E10.3	PROPS(*,2)	Poisson's ratio
	E10.3	PROPS(*,3)	Unit weight
	E10.3	PROPS(*,4)	Compressive strength
	E10.3	FT()	Tensile strength
	E10.3	GF()	Fracture energy
	I5	NLINE()	Tension-softening model indicator (0: bilinear, 1: linear)
Σ MDP (NNCRK)	I7	MOP(*,1,*)	Crack node number (i-tip) of each crack
	I7	MOP(*,2,*)	Crack node number (j-tip) of each crack
	I10	MPMT()	Material number related to crack node number (material number can change along the crack path; if a single material is used, the number is the same.)

(Continued)

Table 9.3 Input Data Formats—cont'd

Number of Lines	Format	Variable	Description	
MINTS	I5	IONTS	Internal spring number	
	I5	NUMIN()	Dummy element number	
	I5	LNOIN(*,1)	Crack node to which internal spring is connected (i-tip)	
	I5	LNOIN(*,2)	Crack node to which internal spring is connected (j-tip)	Not necessary if the crack surface coincides with the boundary as in half-model problems
	I5	ISPID(*,1)	Spring connection condition (in the direction normal to the crack surface)	
	I5	ISPID(*,2)	Spring connection condition (in the direction tangential to the crack surface)	
	I5	MMSTP()	Crack propagation step number (99: notch)	
NNUP	I5	NPUP1	Starting node number of a dummy path for path correction	
	I5	NUMB	Number of nodes on a dummy path for path correction	
1	E10.1	SPRNS(1)	Internal spring coefficient (rigid connection)	
	E10.1	SPRNS(2)	Internal spring coefficient (free surface)	
1	I5	NFYT	Number of nodes on which concentrated load acts as external load	
	I5	NEDGE	Number of element sides on which distributed load acts as external load	
NFYT	I5	NFY()	Node number of a node acted upon by concentrated load	
	E10.3	POINT(*,1)	Unit load component in the horizontal direction	
	E10.3	POINT(*,2)	Unit load component in the vertical direction	
2 × NEDGE	I5	NEASS(*)	Element number of an element acted upon by distributed load	
	I5	NOPRS(*,1)	Node constituting an element side acted upon by distributed load (node i)	
	I5	NOPRS(*,2)	Node constituting an element side acted upon by distributed load (node j)	

Table 9.3 Input Data Formats—cont'd

Number of Lines	Format	Variable	Description	
	E10.3	PUNIT(*,1,1)	Unit load at node *i* in the direction normal to an element side	Order for node numbering
	E10.3	PUNIT(*,1,2)	Unit load at node *i* in the direction tangential to an element side	
	E10.3	PUNIT(*,2,1)	Unit load at node *j* in the direction normal to an element side	
	E10.3	PUNIT(*,2,2)	Unit load at node *j* in the direction tangential to an element side	Positive directions of distributed loads acting on an element side
1	I5	NFOR2	Parameter of constant concentrated load (0: not considered; 1: considered)	
	I5	NEDG2	Parameter of constant distributed load (0: not considered; 1: considered)	
	I5	IGRAV	Parameter for self-weight (0: not considered; 1: considered)	
NFOR2	I5	NFRN()	Node number of a node on which constant concentrated load acts	
	E10.3	FRNX()	X direction component of constant concentrated load	
	E10.3	FRNY()	Y direction component of constant concentrated load	
NEDG2	I5	NEAS2()	Element number of an element on which constant distributed load acts	
	I5	NOPR2(*,1)	Node constituting an element side on which constant distributed load acts (node *i*)	
	I5	NOPR2(*,2)	Node constituting an element side on which constant distributed load acts (node *j*)	
	E10.3	PRN2X(*,1)	Load at node *i* in the direction normal to an element side	
	E10.3	PRN2Y(*,1)	Load at node *i* in the direction tangential to an element side	
	E10.3	PRN2X(*,2)	Load at node *j* in the direction normal to an element side	
	E10.3	PRN2Y(*,2)	Load at node *j* in the direction tangential to an element side	
1	E10.3	THETA	Angle from the *y*-axis along which gravity acts	

9.4 PROGRAM LIST

The section that lists the program (CAIC-M1.FOR) with brief descriptions is available on this book's companion website.

9.5 SELECTED EXAMPLES ILLUSTRATING THE USAGE OF THE PROGRAM

Three sample problems are solved in this section to illustrate the usage of the program.

9.5.1 Crack Analysis of Notched Beam

The solution of a beam problem is presented.

Initial Conditions

A previously studied beam problem, Case 3-4 in Figure 6.1, was selected as the first example. For details of the notched beam and the load condition, refer to Figure 6.1. The material properties used in the analysis are shown in Table 4.1. The bilinear tension-softening model of Figure 4.9 was employed for the crack analysis.

FE Model

Figure 9.12 shows the FE model of the beam. In the numerical analysis the problem was solved under the plane stress condition without taking the self-weight of the beam into consideration.

Input Data

Part of the input data used in the analysis is shown on this book's companion website.

Numerical Results

Load-displacement relations: Figure 9.13 presents the obtained load-midspan displacement relations. Also shown are the results obtained with the plane strain condition. When the beam was analyzed as a plane stress problem, no valid crack propagation pattern was obtained at several postpeak steps, so the computation moved on to the next step of analysis by skipping these no-solution steps.

Crack propagation patterns: Table 9.4 shows the crack propagation patterns obtained at all of the computational steps with the plane stress condition.

Graphic presentation of numerical results: Figure 9.14 presents graphic illustrations of numerical results at three load levels.

Example of output file: Part of the output file is shown on this book's companion website.

9.5.2 Crack Analysis of Scale Model Dam

The solution of a dam problem is presented.

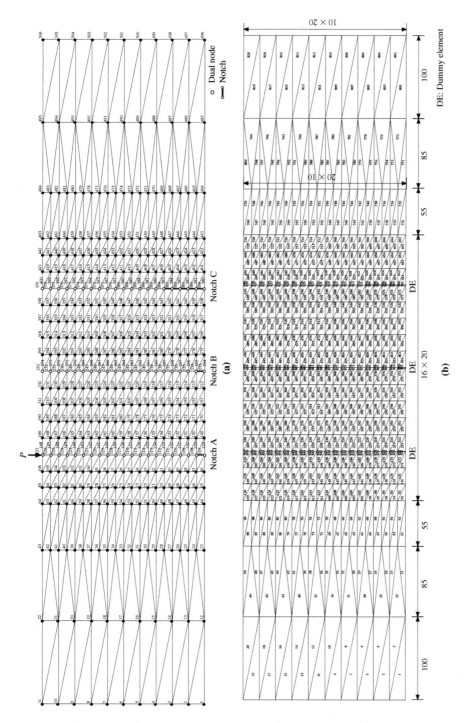

FIGURE 9.12 FE model (dimensions in mm) of notched beam: (a) with node number and (b) with element number. (*Note*: Refer to this book's companion website for an enlarged version of this figure.)

293

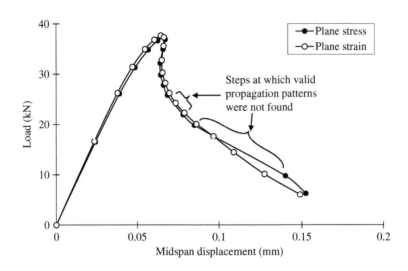

FIGURE 9.13 Obtained load-displacement relations in beam problem.

Table 9.4 Specifications of Output Data on Crack Propagation in the Beam Problem (***.ld file)

Computation Step	Load (kN)	Crack Propagation Step			Crack Depth from the Underside of the Beam (mm)			Remarks on the State of Each Crack
		Crack A	Crack B	Crack C	Crack A	Crack B	Crack C	
1	16.49	0	0	0	10	10	50	Initial state (onset of crack propagation)
2	26.09	1	0	1	20	10	60	A&C: simultaneous propagation
3	31.29	2	1	2	30	20	70	All cracks: simultaneous propagation
4	34.84	3	1	3	40	20	80	A&C: simultaneous propagation; B: stationary
5	36.63	4	1	4	50	20	90	"
6	37.37	5	1	4	60	20	90	A: propagation; B&C: stationary

Table 9.4 Specifications of Output Data on Crack Propagation in the Beam Problem (***.ld file)—cont'd

Computation Step	Load (kN)	Crack Propagation Step			Crack Depth from the Underside of the Beam (mm)			Remarks on the State of Each Crack
		Crack A	Crack B	Crack C	Crack A	Crack B	Crack C	
7	36.88	6	0	4	70	10	90	A: propagation; B: completely closed; C: stationary
8	34.93	7	0	3	80	10	80	A: propagation; C: closing
9	32.14	8	0	3	90	10	80	A: propagation; C: stationary
10	29.79	9	0	2	100	10	70	A: propagation; C: closing
11	27.80	10	0	2	110	10	70	A: propagation; C: stationary
12	25.80	11	0	2	120	10	70	"
13			↓					Jump without finding valid propagation patterns
14	21.94	13	0	1	140	10	60	A: propagation; C: closing
15	19.80	14	0	1	150	10	60	A: propagation; C: stationary
16			↓					Jump without finding valid propagation patterns
17								
18	9.77	17	0	0	180	10	50	A: propagation; C: completely closed
19	6.26	18	0	0	190	10	50	A: propagation (near penetration and finish)

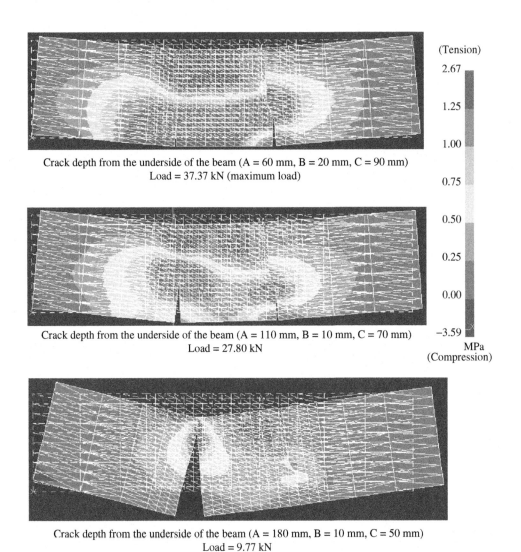

Crack depth from the underside of the beam (A = 60 mm, B = 20 mm, C = 90 mm)
Load = 37.37 kN (maximum load)

Crack depth from the underside of the beam (A = 110 mm, B = 10 mm, C = 70 mm)
Load = 27.80 kN

Crack depth from the underside of the beam (A = 180 mm, B = 10 mm, C = 50 mm)
Load = 9.77 kN

FIGURE 9.14 Graphic illustration of crack propagation, beam deformation, and stress distribution.

Initial Conditions

A previously studied fracture problem of the scale-model dam, Model II-3 in Figure 4.24, was chosen as the second example. For details of the geometric dimensions, notch arrangements, and load conditions, refer to Figures 4.21 and 4.24. Table 4.3 presents the material properties used in the analysis. Notice that the self-weight of the dam was considered in the crack analysis.

FE Model

Figure 9.15 shows the FE model of the dam specimen, with the bottom layers of high-tensile strength to prevent cracking at the base. Crack analysis was carried out under the plane strain condition.

Input Data

Part of the input data used in the analysis is shown on this book's companion website.

Numerical Results

Crack propagation patterns: Table 9.5 summarizes the crack propagation patterns obtained at all the computational steps.

Graphic presentation of numerical results: Figure 9.16 presents graphic illustrations of numerical results at three load levels.

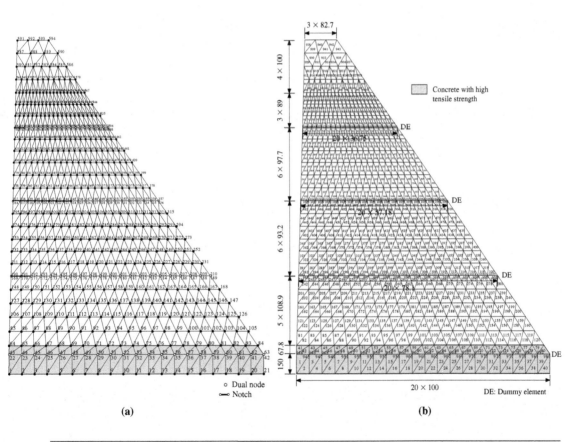

(a)   (b)

FIGURE 9.15 FE model (dimensions in mm) of dam specimen with notches: (a) with node number and (b) with element number. (*Note:* Refer to this book's companion website for an enlarged version of this figure.)

Table 9.5 Specifications of Output Data on Crack Propagation in the Dam Problem (***.ld file)

| Computation Step | Load (kN) | Crack Propagation Step | | | Crack Depth from the Upstream Surface (mm) | | | Remarks on the State of Each Crack |
		Crack A	Crack B	Crack C	Crack A	Crack B	Crack C	
1	487.8	0	0	0	156	457	74	Initial state (onset of crack propagation)
2	676.8	0	1	0	156	515	74	B: propagation
3	685.9	0	2	0	156	572	74	"
4	654.9	0	3	0	156	629	74	"
5	623.9	0	4	0	156	686	74	"
6	538.7	0	5	0	156	743	74	"
7	518.8	0	6	0	156	801	74	"
8	431.0	0	7	0	156	858	74	"
9	340.8	0	8	0	156	915	74	"
10	262.4	0	9	0	156	972	74	"
11	196.2	0	10	0	156	1029	74	"
12	140.9	0	11	0	156	1086	74	B: propagation (near penetration and finish)

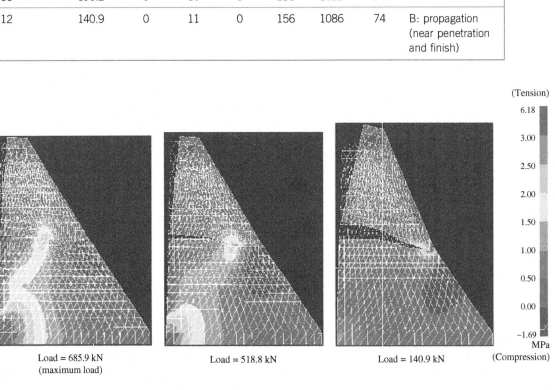

Load = 685.9 kN (maximum load)   Load = 518.8 kN   Load = 140.9 kN

(Tension)
6.18
3.00
2.50
2.00
1.50
1.00
0.50
0.00
−1.69
MPa
(Compression)

FIGURE 9.16 Graphic illustration of crack propagation, dam deformation, and stress distribution.

9.5.3 Crack Analysis of Tunnel Lining

The solution of a tunnel problem is presented.

Initial Conditions

A previously studied tunnel lining problem, the half-model of the tunnel specimen in Figure 4.17, was selected as the third example. For the geometric details, notch arrangements, and load conditions, refer to Figures 4.15 and 4.17. The material properties used in numerical studies are shown in Table 4.2.

FE Model

Figure 9.17 presents a half-FE-model for the tunnel specimen. Crack analysis was carried out under the plane strain condition. In the analysis, crack paths were fixed, and the self-weight of the specimen was not taken into account (to reflect the test condition).

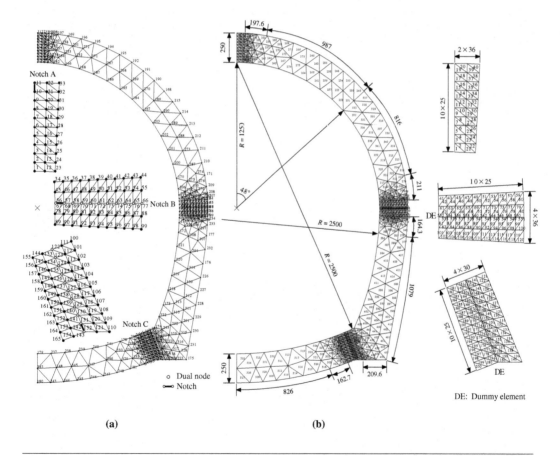

(a)   (b)

FIGURE 9.17 Half-FE-model (dimensions in mm) of tunnel specimen with notches: (a) with node number and (b) with element number. (*Note:* Refer to this book's companion website for an enlarged version of this figure.)

Notice that the strength tolerance parameter, AAA, was assumed to be 1.10 in the crack analysis.

Input Data

Part of the input data used in the analysis is shown on this book's companion website.

Numerical Results

Crack propagation patterns: Table 9.6 shows the crack propagation patterns obtained at all the computational steps.

Graphic presentation of numerical results: Figure 9.18 presents graphic illustrations of numerical results at three load levels.

Table 9.6 Specifications of Output Data on Crack Propagation in the Tunnel Problem (***.ld file)

Computation Step	Load (N/mm^2)	Crack Propagation Step			Crack Depth (mm)			Remarks on the State of Each Crack
		Crack A	Crack B	Crack C	Crack A	Crack B	Crack C	
1	0.0397	0	0	0	25	25	25	Initial state (onset of crack propagation)
2	0.0588	0	1	1	25	50	50	A: stationary; B&C: propagation
3	0.0713	1	2	2	50	75	75	All cracks: propagation
4	0.0726	1	3	3	50	100	100	A: stationary; B&C: propagation
5	0.0680	1	4	3	50	125	100	A&C: stationary; B: propagation
6	0.0642	1	5	3	50	150	100	"
7	0.0613	1	6	4	50	175	125	A: stationary; B&C: propagation
8	0.0553	2	7	5	75	200	150	All cracks: propagation (near penetration and finish)

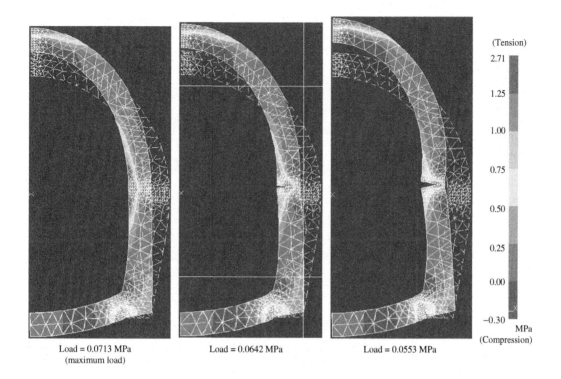

Load = 0.0713 MPa   Load = 0.0642 MPa   Load = 0.0553 MPa
(maximum load)

FIGURE 9.18 Graphic illustration of crack propagation, tunnel deformation, and stress distribution.

REFERENCE

Owen, D.R.J., and Hinton E. (1980). *Finite Elements in Plasticity: Theory and Practice.* Pineridge Press Limited.

Computer Program for Mixed-Mode Type Crack Analysis in Concrete Using EFCM (CAIC-M12.FOR)

10

10.1 OVERVIEW OF THE PROGRAM

This program, CAIC-M12.FOR, which has been developed by extending the program CAIC-M1.FOR introduced in the preceding chapter, performs crack analysis of the mixed-mode type (mode I + mode II) in structural concrete. Except for the fracture mode involved, the main features of the program are basically the same as those of CAIC-M1.FOR. Table 10.1 lists the subroutines that have been added in CAIC-M12.FOR for mixed-mode crack analysis.

10.2 STRUCTURE OF THE PROGRAM

Basically, the structure and flow of this program are the same as those of the mode-I crack analysis program, CAIC-M1.FOR, and the flowcharts of Figure 9.1 and Figure 9.2 shown in Chapter 9 are applicable to this program. Solving a mixed-mode crack problem, however, requires additional operations of setting pairs of unit cohesive forces in the direction tangential to the crack surface, calculating sliding displacement and other coupled influence coefficients, and formulating the crack equation for mixed-mode fracture, as shown in Figure 10.1.

10.3 MAIN RULES

Main rules and issues related to the programming are discussed in this section.

Shear-Transfer Modeling

Figure 10.2 presents two types of shear-transfer model employed in the program—the bilinear type and the trilinear type—and the variables used to define these models. The trilinear model also enables quadrilateral and trapezoidal modeling, as shown in Figure 10.3. In this program, after reaching the shear strength at $SW1*W_c$, the shear-transfer process is treated as irreversible, similar to the tension-softening modeling.

Table 10.1 Additional Subroutines for Mixed-Mode Crack Analysis

Name of Subroutine	Main Function
COEFS	Setting a unit cohesive force in the direction tangential to the crack surface
FSTIP	Calculating nodal force in the direction tangential to the crack surface

Formulation of Crack Equation

Taking the problem of Figure 7.8 as an example, this section explains the formulation of the crack equation for mixed-mode fracture in matrix form.

Conditions of the problem (Figure 7.8):
   (a) Fracture mode: Mixed mode (mode I + mode II)
   (b) Number of cracks: 2 (crack A and crack B)
   (c) State of crack propagation (number of crack nodes):
      Crack A: crack node $i = 1$ *to* N [NR(A) = N in TFORCE]
      Crack B: crack node $j = 1$ *to* M [NR(B) = M in TFORCE]
   (d) Main crack: Crack A

Notice that physical quantities related to an inactive crack (crack B) are denoted with asterisks.
   The crack equation is expressed in matrix form as

$$[SD] = [SE] \times [SF] \tag{10.1}$$

with

$$[SD]^T = \left[BKN_n^a \ Q_n^a \ BKN_t^a \ Q_t^a \ BKN_n^b \ Q_n^b \ BKN_t^b \ Q_t^b, -(Q_{nl}^a + CRN_n^a) \right] \tag{10.2}$$

$$[SE] = \begin{bmatrix}
E_N & -AK_{nn}^{aa} & 0 & -AK_{nt}^{aa} & 0 & -AK_{nn}^{ab} & 0 & -AK_{nt}^{ab} & -BK_n^a \\
S_n^a & T \cdot E_N & 0 & 0 & 0 & 0 & 0 & 0 & 0 \\
0 & -AK_{tn}^{aa} & E_N & -AK_{tt}^{aa} & 0 & -AK_{tn}^{ab} & 0 & -AK_{tt}^{ab} & -BK_t^a \\
S_t^a & 0 & 0 & T \cdot E_N & 0 & 0 & 0 & 0 & 0 \\
0 & -AK_{nn}^{ba} & 0 & -AK_{nt}^{ba} & E_M & -AK_{nn}^{bb} & 0 & -AK_{nt}^{bb} & -BK_n^b \\
0 & 0 & 0 & 0 & S_n^b & T \cdot E_M & 0 & 0 & 0 \\
0 & -AK_{tn}^{ba} & 0 & -AK_{tt}^{ba} & 0 & -AK_{tn}^{bb} & E_M & -AK_{tt}^{bb} & -BK_t^b \\
0 & 0 & 0 & 0 & S_t^b & 0 & 0 & T \cdot E_M & 0 \\
0 & CI_{nn}^{aa} & 0 & CI_{nt}^{aa} & 0 & CI_{nn}^{ab} & 0 & CI_{nt}^{ab} & CR_n^a
\end{bmatrix} \tag{10.3}$$

$$[SF]^T = \left[W_n^a \ F_n^a \ W_t^a \ F_t^a \ W_n^{b*} \ F_n^{b*} \ W_t^{b*} \ F_t^{b*} \ P^a \right] \tag{10.4}$$

where superscripts a and b represent crack A and crack B, respectively; subscript n denotes a physical quantity in the direction perpendicular to the crack surface, and subscript t stands for a physical quantity in the direction tangential to the crack surface. The definitions of the matrices and variables are given in the list that begins on page 306.

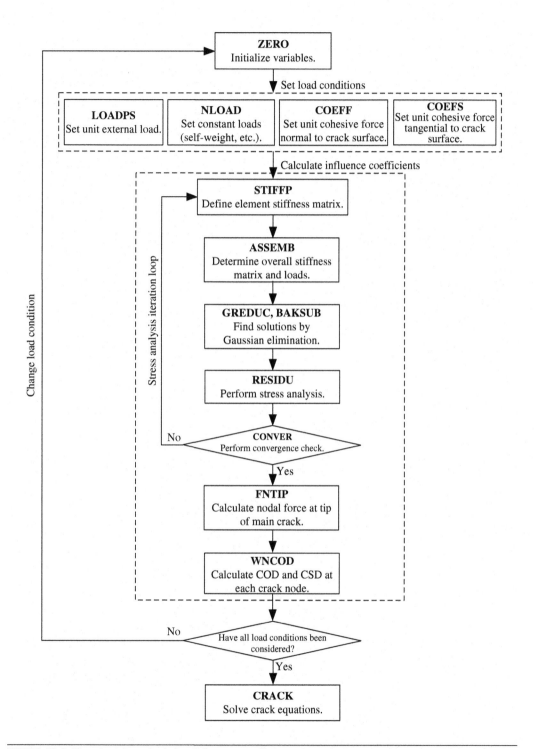

FIGURE 10.1 Flowchart for calculating influence coefficients and solving crack equations in mixed-mode crack analysis (EFFECT).

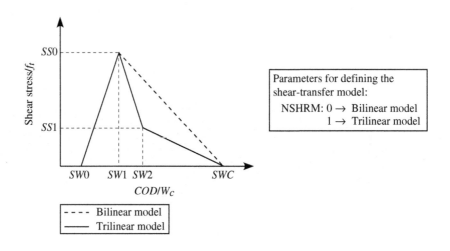

FIGURE 10.2 Definitions and variables used in shear-transfer modeling

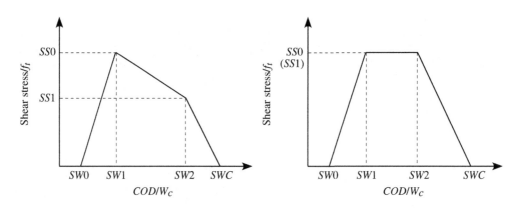

FIGURE 10.3 Examples of the trilinear type in shear-transfer modeling.

[SD]—Vector of strength properties:

BKN_n^a, BKN_n^b: Column vectors of the order N and M representing the crack opening displacement due to constant loads at each node of crack A and crack B, respectively

BKN_t^a, BKN_t^b: Column vectors of the order N and M representing the crack sliding displacement due to constant loads at each node of crack A and crack B, respectively

Q_n^a, Q_n^b: Column vectors of the order N and M representing the Y-intercept in the tension-softening model at each node of crack A and crack B, respectively

Q_t^a, Q_t^b: Column vectors of the order N and M representing the Y-intercept in the shear-transfer model at each node of crack A and crack B, respectively

Q_{nl}^a: Limit nodal force at the tip of crack A

CRN_n^a: Nodal force component at the tip of crack A due to constant loads

[SE]—Influence coefficient matrix:

E_N, E_M: Unit matrices of the order (N, N) and (M, M) respectively

S_n^a, S_n^b: Matrices of the order (N, N) and (M, M) representing the slope in the tension-softening model corresponding to the crack opening displacement of W_n^a and W_n^{b*}, respectively

S_t^a, S_t^b: Matrices of the order (N, N) and (M, M) representing the slope in the shear-transfer model corresponding to the crack opening displacement of W_n^a and W_n^{b*}, respectively

Examples of influence coefficients in the mixed-mode formulation are shown in the following equation.

$$AK_{nt}^{ab} = \begin{bmatrix} AK_{nt}^{ab11} & AK_{nt}^{ab12} & \cdot & \cdot & AK_{nt}^{ab1M} \\ AK_{nt}^{ab21} & AK_{nt}^{ab22} & \cdot & \cdot & AK_{nt}^{ab2M} \\ \cdot & & \cdot & \cdot & \cdot \\ \cdot & & & \cdot & \cdot \\ AK_{nt}^{abN1} & AK_{nt}^{abN2} & \cdot & \cdot & AK_{nt}^{abNM} \end{bmatrix} \tag{10.5}$$

Matrix of the order (N, M) representing the crack opening displacement at each node of crack A due to a unit shear force applied at the nodes of crack B

Table 10.2 lists all the influence coefficients that are defined in the crack equation for mixed-mode crack analysis. These influence coefficients are stored by using the following variables in the program:

$$AK(N+M, N+M) = \begin{bmatrix} AK_{nn}^{aa} & AK_{nn}^{ab} \\ AK_{nn}^{ba} & AK_{nn}^{bb} \end{bmatrix} \tag{10.6}$$

$$AKNT(N+M, N+M) = \begin{bmatrix} AK_{nt}^{aa} & AK_{nt}^{ab} \\ AK_{nt}^{ba} & AK_{nt}^{bb} \end{bmatrix} \tag{10.7}$$

$$AKTN(N+M, N+M) = \begin{bmatrix} AK_{tn}^{aa} & AK_{tn}^{ab} \\ AK_{tn}^{ba} & AK_{tn}^{bb} \end{bmatrix} \tag{10.8}$$

$$AKTT(N+M, N+M) = \begin{bmatrix} AK_{tt}^{aa} & AK_{tt}^{ab} \\ AK_{tt}^{ba} & AK_{tt}^{bb} \end{bmatrix} \tag{10.9}$$

BK_n^a, BK_n^b: Column vectors of the order N and M representing the crack opening displacement under a unit external load at each node of crack A and crack B, respectively

BK_t^a, BK_t^b: Column vectors of the order N and M representing the crack sliding displacement under a unit external load at each node of crack A and crack B, respectively

CI_{nn}^{aa}, CI_{nn}^{ab}: Row vectors of the order N and M representing the nodal force component at the tip of crack A due to a unit normal force applied at the nodes of crack A and crack B, respectively

Table 10.2 Definition of Influence Coefficients in the Crack Equation for Mixed-Mode Crack Analysis

Influence Coefficient	Physical Meaning	Calculated at	Direction of Unit Cohesive Force	Applied at
AK_{nn}^{aaik}	Resulting COD	i-node of crack A	Perpendicular to the crack surface	k-node of crack A
AK_{nn}^{abij}	"	"	"	j-node of crack B
AK_{nn}^{baji}	"	j-node of crack B	"	i-node of crack A
AK_{nn}^{bbjk}	"	"	"	k-node of crack B
AK_{nt}^{aaik}	"	i-node of crack A	Tangential to the crack surface	k-node of crack A
AK_{nt}^{abij}	"	"	"	j-node of crack B
AK_{nt}^{baji}	"	j-node of crack B	"	i-node of crack A
AK_{nt}^{bbjk}	"	"	"	k-node of crack B
AK_{tn}^{aaik}	Resulting CSD	i-node of crack A	Perpendicular to the crack surface	k-node of crack A
AK_{tn}^{abij}	"	"	"	j-node of crack B
AK_{tn}^{baji}	"	j-node of crack B	"	i-node of crack A
AK_{tn}^{bbjk}	"	"	"	k-node of crack B
AK_{tt}^{aaik}	"	i-node of crack A	Tangential to the crack surface	k-node of crack A
AK_{tt}^{abij}	"	"	"	j-node of crack B
AK_{tt}^{baji}	"	j-node of crack B	"	i-node of crack A
AK_{tt}^{bbjk}	"	"	"	k-node of crack B

$CI_{nt}^{aa}, CI_{nt}^{ab}$: Row vectors of the order N and M representing the nodal force component at the tip of crack A due to a unit shear force applied at the nodes of crack A and crack B, respectively

CR_n^a: Nodal force at the tip of crack A due to a unit external load (< 0)

T: Member thickness

[SF]—Vector of unknowns:

W_n^a, W_n^{b*}: Column vectors of the order N and M representing the crack opening displacement at each node of crack A and crack B, respectively

W_t^a, W_t^{b*}: Column vectors of the order N and M representing the crack sliding displacement at each node of crack A and crack B, respectively

F_n^a, F_n^{b*}: Column vectors of the order N and M representing the cohesive force normal to the crack surface at each node of crack A and crack B, respectively

F_t^a, F_t^{b*}: Column vectors of the order N and M representing the shear force at each node of crack A and crack B, respectively

P^a: External load required to propagate crack A

Signs of Crack Sliding Displacement and Shear Force

In establishing the crack equation, the direction of shear force must be carefully determined, based on the direction of crack surface sliding. To achieve consistency with the signs of the crack sliding displacement calculated by WNCOD, the right-to-left sliding direction (when viewing the crack surface from outside) as shown in Figure 10.4 is defined as positive.

The three steps for determining the direction of shear force on a crack surface are as follows.

1. As an influence coefficient due to a unit external load, the crack sliding displacement at the first node of a crack (the notch tip) is obtained in EFFECT [the name of the variable in the program: BKS(1, ICRAK); ICRAK: the crack of interest].

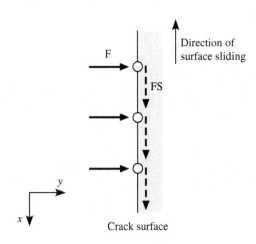

F : Cohesive force normal to crack surface
FS: Shear force
O : Dual node (i-tip or j-tip)

FIGURE 10.4 Basic rules for determining the directions of crack surface sliding and cohesive forces.

2. If BKS(1, ICRAK) $\geq$ 0, shear forces should act in the direction shown in Figure 10.4. If BKS(1, ICRAK) < 0, they should be applied in the opposite direction on the crack surface.

3. When determining the shear-transfer characteristics in the matrices [SD] and [SE] of the crack equation in CRACK, the first two rules are applied to the signs of Q_t and S_t.

Input Data

Table 10.3 shows the additional input data formats that are required for this program in addition to those shown in Table 9.3 for mode-I crack analysis. The only changes from Table 9.3 are related to material properties: the shear strength and the shear-transfer model.

Table 10.3 Additional Input Data Formats (Only Changes from Table 9.3)

Number of Lines	Format	Variable	Description
NMATS (number of material types)	E10.3	PROPS (*,1)	Modulus of elasticity
	E10.3	PROPS (*,2)	Poisson's ratio
	E10.3	PROPS (*,3)	Unit weight
	E10.3	PROPS (*,4)	Compressive strength
	E10.3	FT()	Tensile strength
	E10.3	GF()	Fracture energy
	I5	NLINE()	Tension-softening model indicator (0: bilinear; 1: linear)
	E10.3	SS0()	Ratio of shear strength to tensile strength
	I5	NSHRM	Shear-transfer model indicator (0: bilinear; 1: trilinear)
NMATS (number of material types)	E10.3	SS1()	Parameter for defining the shear-transfer model [ratio to SS0(); refer to Figure 10.2.]
	E10.3	SW0()	Parameters for defining the shear-transfer model (ratio to W_c; refer to Figure 10.2); For bilinear models, only SW0, SW1 and SWC are required
	E10.3	SW1()	
	E10.3	SW2()	
	E10.3	SWC()	

10.4 SUBROUTINES WITH MAJOR CHANGES

Subroutines with major changes in CAIC-M12.FOR are explained in this section.

10.4.1 Changes in CAIC-M12.FOR from CAIC-M1.FOR

Figure 10.5 lists the added or modified subroutines in CAIC-M12.FOR. In the following, the subroutines in the boldfaced boxes are briefly explained, which constitute the major changes from CAIC-M1.FOR. The remaining subroutines are not described because they contain only minor changes related to the newly added variables and can be easily understood. The newly added variables are summarized in Table 10.4.

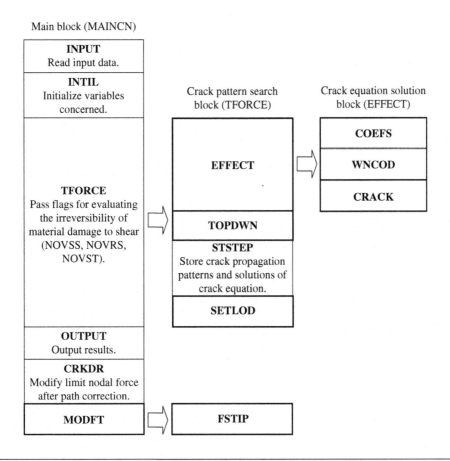

FIGURE 10.5 Subroutines added or modified in CAIC-M12.FOR.

Table 10.4 New Key Variables in CAIC-M12.FOR

Variable		Description
Related to material properties	QFS(*,*)	Nodal force equivalent to shear strength
	FSMIN(*,*)	Minimum value of the shear force transferred so far at each crack node
	SS()	Shear strength
	SS0() SS1() SW0() SW1() SW2() SWC()	See Figures 10.2 and 10.3
	NSHRM	Parameter for defining the shear-transfer model (0: bilinear; 1: trilinear)
Related to crack equation	Solution	FS(*,*)	Shear force at each crack node in the true crack propagation pattern
		WS(*,*)	Sliding displacement at each crack node in the true crack propagation pattern
		FSS(*,*)	Shear force at each crack node in each crack propagation pattern
		WWS(*,*)	Sliding displacement at each crack node in each crack propagation pattern
	Influence coefficient	SUBER	Sliding displacement at a crack node
		BKS(*,*)	Sliding displacement at each crack node due to a unit external load
		BKSN(*,*)	Sliding displacement at each crack node due to a constant load
		CIS(*,*)	Nodal force component at a crack tip due to a unit shear force
		AKTN(*,*)	Sliding displacement at each crack node due to a unit cohesive force in the direction perpendicular to the crack surface
		AKNT(*,*)	Crack opening displacement at each crack node due to a unit shear force
		AKTT(*,*)	Sliding displacement at each crack node due to a unit shear force
Related to flags		NOVSS(*,*) NOVRS(*,*) NOVST(*,*)	Flag for judging whether or not the sliding displacement at a crack node exceeds SW0(NMATS) (0: not exceeded; 1: exceeded); If SW0(NMATS) is exceeded, the irreversibility of material damage to shear is taken into consideration

10.4.2 Subroutines with Major Changes in the Crack Pattern Determination Block (TFORCE)

For information about the code for the subroutines, please see this book's companion website.

10.4.3 Subroutines with Major Changes in the Crack Equation Solution Block (EFFECT)

For information about the code for the subroutines, please see this book's companion website.

10.4.4 Subroutines with Major Changes in the Main Block (MAINCN)

For information about the code for the subroutines, please see this book's companion website.

10.5 SELECTED EXAMPLE ILLUSTRATING THE USAGE OF THE PROGRAM

A sample problem is solved in this section to illustrate the usage of the program.

Initial Conditions

A previously studied mixed-mode fracture problem—the single-notched shear beam shown in Figure 7.10—was selected as an example to illustrate the usage of the program. The material properties used in the analysis are shown in Table 7.1. For tension-softening the bilinear model of Figure 7.4 was used, and for shear transfer the bilinear model of Case 1 in Figure 7.11 was employed. The problem was solved under the plane stress condition, without taking the self-weight into consideration.

FE Model

Figures 10.6a and 10.6b show the FE model of the beam. As seen from the figures, the preset initial crack path with dual nodes has a steep turn at the notch tip, and the remaining path coincides with a dummy path designated for path correction. To avoid changing the basic rules for path correction and to simplify programming, the first-step crack propagation was defined based on the analytically predetermined direction of principal stress at the tip of the notch, and in MAINCN before entering CRKDR for path correction the next dummy path was reassigned. Due to this irregularity in presetting the crack path in the present problem, some modifications were made in the program regarding the calculation of the limit nodal force at the notch tip and the calculation of cohesive forces. These changes are noted in the comments of the attached program.

Input Data

Part of the input data used in the analysis is shown on this book's companion website.

Numerical Results

Figure 10.7 presents graphic illustrations of numerical results at three load levels. Tensile stresses exceeding the tensile strength ($3.57 N/mm^2$) were observed in the vicinity of the crack path. This was believed to be caused by the discrepancy between the shear-transfer model used in the crack analysis and the actual shear-transfer characteristics in the real material.

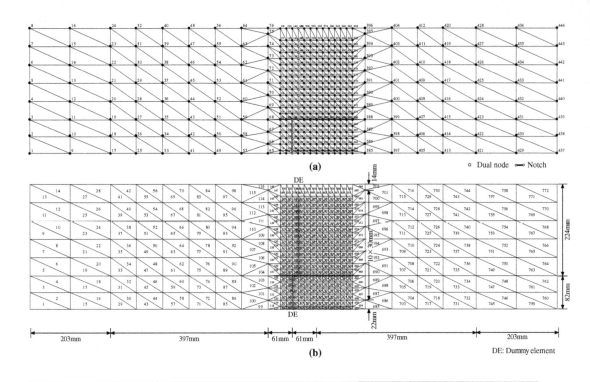

FIGURE 10.6A FE model: (a) with node number and (b) with element number. (*Note:* Refer to this book's companion website for an enlarged version of this figure.)

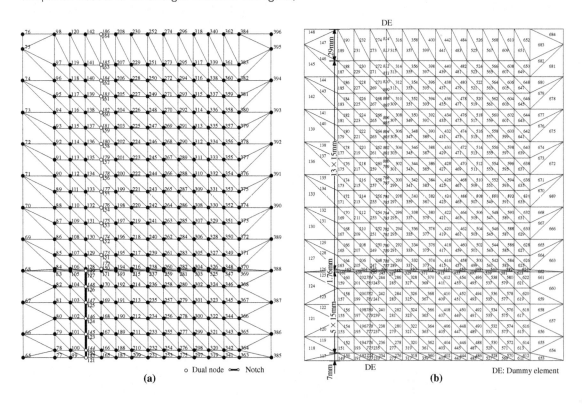

FIGURE 10.6B Partially enlarged view of the FE models: (a) with node number and (b) with element number. (*Note*: Refer to this book's companion website for an enlarged version of this figure.)

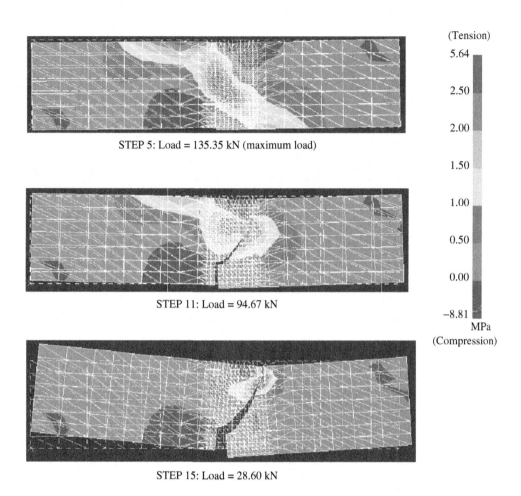

STEP 5: Load = 135.35 kN (maximum load)

STEP 11: Load = 94.67 kN

STEP 15: Load = 28.60 kN

(Tension)
5.64
2.50
2.00
1.50
1.00
0.50
0.00
−8.81
MPa
(Compression)

FIGURE 10.7 Graphic illustrations of crack propagation, beam deformation, and stress distribution.

INDEX

A

AAA variable, 280–281
ACI 446 report, 5
Adachi-Oka model, 228–231
Airy stress function
  elastic crack-tip fields, 25–29
  plane elastic problems, 30–31
AK variable, 284
Alternative loadings, with notched beams, 159–163
Arch cracks, 245, 250, 253
Arrea, M., single-notched shear beam tests, 181–182
  five shear-COD relations, 193–197
  three shear-COD relations, 197–206

B

Backfilling voids, 104–105
Barenblatt, G. I.
  cohesive zone models, 45–47
  FCM, 5–6
Bazant, Z. P.
  FCM modeling, 68
  microplane theory, 4
  mixed-mode problems, 17, 179–181
  particle model, 4
Beams
  under bending, 92
    curvilinear crack paths, 97–101
    FCM, 7
    fixed crack path, 92–96
  EFCM, 82
  notched. *See* Notched beams
  tunnel lining models, 216, 219
Bending
  notched beams. *See* Notched beams
  simple beams under, 92
    curvilinear crack paths, 97–101
    FCM, 7
    fixed crack path, 92–96
Best-fit curves, in uniaxial tension tests, 56–57
Biharmonic equations, 25, 27

Bilinear shear
  CAIC-M12.FOR program, 303
  gravity dams, 207–208
  single-notched beams, 193–194, 202
Bilinear tension-softening
  fixed crack path analysis, 92–94
  FPZ cohesive forces, 183–184
  gravity dams, 111, 207
  ground pressure database, 248
  inverse modeling method, 57, 60–62
  notched beams, 128, 152, 193
  tunnel lining models, 222, 231
BK variable
  CAIC-M1.FOR program, 284
  CAIC-M12.FOR program, 307
BKN variable, 306
BKS variable, 309–310
Boundary conditions
  multiple-crack problems, 15
  tunnel lining models, 215
Brittle fracture, 5

C

CAIC-M1.FOR program, 271
  crack equations, 283–285
  crack path correction, 286–288
  crack path modeling, 278–279
  crack propagation state variables, 282–283
  input data, 287–291
  list, 292
  main cracks and subcracks, 279
  main flags, 281–282
  multiple-crack analysis, 280–281
  nodal forces, 279
  notched beams, 292–296
  scale model dam, 292, 296–298
  self-weight and other constant loads, 286
  signs in, 285
  structure, 271–277
  termination of computation, 286
  triangular elements, 278
  tunnel lining, 299–301
  validity of solutions, 279

317

Printed in the United States
By Bookmasters